ST/ESA/STAT/SER.W/26

Department of Economic and Social Affairs
Statistics Division

I0091717

2014
Electricity Profiles

United Nations
New York, 2016

The Department of Economic and Social Affairs of the United Nations Secretariat is a vital interface between global policies in the economic, social and environmental spheres and national action. The Department works in three main interlinked areas: (i) it compiles, generates and analyses a wide range of economic, social and environmental data and information on which States Members of the United Nations draw to review common problems and to take stock of policy options; (ii) it facilitates the negotiations of Member States in many intergovernmental bodies on joint courses of action to address ongoing or emerging global challenges; and (iii) it advises interested Governments on the ways and means of translating policy frameworks developed in United Nations conferences and summits into programmes at the country level and, through technical assistance, helps build national capacities.

NOTE

Symbols of United Nations documents are composed of capital letters combined with figures.

General Disclaimer

ST/ESA/STAT/SER.W/26

UNITED NATIONS PUBLICATION
Sales number: E.17.XVII.7

ISBN 978-92-1-161618-7
eISBN 978-92-1-060080-4
ISSN 0257-7208

CONTENTS

Page

CONTENTS

Electricity Profiles (2009 - 2014) (continued)

CONTENTS

Electricity Profiles (2009 - 2014) (continued)

INTRODUCTION

The *Electricity Profiles 2014* provides detailed information on production, trade and consumption of electricity, on net installed capacity and thermal power plant inputs and efficiency for 228 countries and areas on an internationally comparable basis, for the years 2009-2014. It is published by the United Nations Statistics Division with the aim of providing an overall picture of the electricity sector of such countries and areas.

This is the third issue of *Electricity Profiles* as a stand-alone publication, whereas until the 2011 edition it was part of the series *Energy Balances and Electricity Profiles*. The series was split in two and in addition to the *Electricity Profiles* a sister publication, the *Energy Balances*, is released.

This split followed the incorporation of the standards brought about by the *International Recommendations for Energy Statistics*[1] (IRES). IRES, in its draft form, was endorsed in 2011 by the United Nations Statistical Commission. Previously, energy statistics published by the Statistics Division followed the format described in detail in the technical report entitled *Concepts and Methods in Energy Statistics, with Special Reference to Energy Accounts and Balances*[2] and also discussed in the publication *Energy Statistics: A Manual for Developing Countries*.[3]

Electricity production and electric installed capacity are disaggregated by source, as coming from Combustible fuels, Hydro, Nuclear and Other sources; and by type of producer, whether from main activity producers or autoproducers. The latter is given implicitly as the difference between Total production/capacity and Main activity production/capacity.

Thermal power plant inputs are listed by energy product, together with the total output, which are used together to calculate the overall efficiency of electricity production from combustible fuels (displayed at the bottom). Up to ten main products (by input contribution in the latest year available) are displayed and then listed alphabetically, with the remaining ones, if any, aggregated in the category Others.

More detailed disaggregation of electricity capacity and production is published in the *Energy Statistics Yearbook*, where wind and solar are identified separately as electricity sources rather than in the category Other.

The information contained in this publication is also available in electronic format.[4] Requests for information should be directed to United Nations Publications at: order@un.org.

Acknowledgement is due to the following specialized and intergovernmental agencies whose publications have been utilized in supplementing our statistics: Food and Agriculture Organization of the United Nations (FAO), International Atomic Energy Agency (IAEA), International Energy Agency of the Organisation for Economic Cooperation and Development (IEA/OECD), International Sugar Organization (ISO), Organization of Arab Petroleum Exporting Countries (OAPEC), Organization of the Petroleum Exporting Countries (OPEC), Organización Latinoamericana de Energía (OLADE), Statistical Office of the European Union (Eurostat), World Energy Council (WEC). Acknowledgement is also made to governmental, energy and statistical authorities of the Member States which have been cooperative in providing data.

The annual energy data are being collected and processed by the Industrial and Energy Statistics Section of UNSD, headed by Mr. Ralf Becker. The processing of the energy data and preparation for publication were carried out by Mr. Leonardo Souza, Mr. Alexander Blackburn, Mr. Man Soni, Ms. Costanza Giovannelli, Ms. Peng Guo and Mr. Graham Osborn.

Enquiries, comments and suggestions for improving this publication are welcome and should be addressed to: energy_stat@un.org.

---

[1] Available at http://unstats.un.org/unsd/energy/ires
[2] Statistical Papers, Series F, No. 29 (United Nations publication, Sales No.E.82.XVII.13).
[3] Statistical Papers, Series F, No. 56 (United Nations publication, Sales No.E.91.XVII.10).
[4] For details, see http://unstats.un.org/unsd/energy/EProfiles

CONCEPTS AND DEFINITIONS

ELECTRICITY PROFILES

The electricity profiles intend to provide an overall picture of the electricity sector in countries and areas on a comparable basis. As such, a simplified approach is taken, where for production and capacity the three main electricity sources (or production processes) are singled out; namely, combustible fuels, hydro and nuclear; while the remaining electricity production processes (mostly renewable) are aggregated in the category Other. Detailed information on each source for the same countries and areas is available from the United Nations Energy Statistics Database.[5]

Each electricity profile is divided into three blocks, which are listed below and described in detail next:
• The top block on electricity production, trade and consumption;
• The middle block on net installed capacity; and
• The bottom block on combustible fuel input, output and efficiency.

Top block – Electricity production, trade and consumption

Electricity is defined in energy statistics as the transfer of energy through the physical phenomena involving electric charges and their effects when at rest and in motion.

Electricity can be generated through different processes such as: the conversion of energy contained in falling or streaming water, wind or waves; the direct conversion of solar radiation through photovoltaic processes in semiconductor devices (solar cells); or by the combustion of fuels.

Electricity **production** (or electricity generation) in the electricity profiles refers to gross production, which is the sum of the electrical energy production by all the generating units/installations concerned (including pumped storage) measured at the output terminals of the main generators.

The different types of technology/processes for the generation of electricity are defined as follows.

Electricity from **combustible fuels** refers to the production of electricity from the combustion of fuels which are capable of igniting or burning, i.e. reacting with oxygen to produce a significant rise in temperature.

Hydro electricity refers to electricity produced from devices driven by fresh, flowing or falling water.

Nuclear electricity refers to electricity generated by nuclear plants.

Other electricity includes all other processes, which are described below.

• Solar Electricity refers to electricity produced from solar photovoltaics, i.e. by the direct conversion of solar radiation through photovoltaic processes in semiconductor devices (solar cells), including concentrating photovoltaic systems; as well as electricity produced from solar thermal heat (both concentrating and non-concentrating).

Heat from concentrating solar thermal refers to high temperature heat produced from solar radiation captured by concentrating solar thermal systems. Heat from non-concentrating solar thermal refers to low temperature heat produced from solar radiation captured by non-concentrating solar thermal systems.

• Wind electricity refers to electricity produced from devices driven by wind.

• Wave electricity refers to electricity produced from devices driven by the motion of waves.

• Tidal electricity refers to electricity generated from devices driven by tidal currents or the differences of water level caused by tides.

• Other marine electricity refers to electricity generated from devices which exploit sources of marine energy not elsewhere specified. Examples of sources are non-tidal currents, temperature differences and salinity gradients in seas or salinity differences between sea and fresh water.

• Geothermal electricity refers to the electricity generated from the heat from geothermal sources.

• Electricity generated from chemical heat, which refers to recovered heat generated in the chemical industry by exothermic reactions other than combustion.

---

[5] http://data.un.org/Explorer.aspx?d=EDATA

CONCEPTS AND DEFINITIONS

• Electricity from other sources not elsewhere specified.

As regards the type of producer, both for electricity production and for net installed capacity, a distinction is made between main activity producers and autoproducers, as defined below:

Main activity producer are enterprises which produce electricity as their principal activity. Formerly known as public utilities, these enterprises may be privately or publicly owned companies.

Autoproducers are enterprises which produce electricity but for whom the production is not their principal activity.

In the *Electricity Profiles*, autoproducer figures are not provided explicitly, but can be easily derived as the difference between total producers and main activity producers.

Own use in electricity, CHP and heat plants refers to consumption of electricity for the direct support of electricity, CHP and heat plants. It includes consumption by station auxiliaries, and losses in transformers which are considered as integral parts of the electric energy, CHP or heat generating plants.

• **Electricity plants** refer to plants producing only electricity.

• **CHP plants** (Combined Heat and Power) refer to plants which produce both heat and electricity from at least one generating unit in the plant. They are sometimes referred to as "co-generation" plants.

• **Heat plants** refer to plants (including heat pumps and electric boilers) designed to produce heat only, for deliveries to third parties.

Net production is (gross) production minus own use in electricity, CHP and heat plants. It is equivalent to the electricity sent out from the plants that is available to the network or ready for use in the premises in the case of autoproducers.

Imports comprise all electricity entering the national territory.

Exports comprise all electricity leaving the national territory.

For electricity, as an exception to general energy statistics, trade data include electricity transmitted through the country from one neighbour to another, as there is no practical way of discerning which quantities are re-exported and which are consumed by the transit country.

Losses refer to losses during the transmission and distribution of electricity. Losses also include pilferage of electricity (sometimes referred to as non-technical losses).

Consumption in the *Electricity Profiles* refers to **Energy industries own use** and **Final consumption**.

Energy industries own use refers to consumption of electricity for the direct support of the production, and preparation for use of fuels and energy. Quantities which are used for transportation purposes in the energy industry are not included here, but in Consumption by transport. Quantities used for pumped storage are included here.

Note that pumped storage plants are plants where electricity is used during periods of lower demand to pump water into reservoirs for subsequent release and electricity generation during periods of higher demand. Less electricity is eventually produced than is consumed to pump the water into the higher reservoir.

In general energy statistics, Own use in electricity, CHP and heat plants is part of Energy industries own use, but it is displayed apart in this publication so that Net production can be derived.

Final consumption refers to the consumption of electricity by Manufacturing, construction and non-fuel mining, by Transport, and by households and other consumers (non-energy use being irrelevant for electricity).

By industry and construction refers to final electricity consumption by manufacturing, construction and non-fuel mining industries. The final consumption recorded under this category covers the use of electricity by economic units belonging to the industry groups listed below (excluding the use of energy products for transport, which is recorded under Transport in its respective row).
• Iron and steel
• Chemical and petrochemical
• Non-ferrous metals

CONCEPTS AND DEFINITIONS

- Non-metallic minerals
- Transport equipment
- Machinery
- Mining and quarrying
- Food and tobacco
- Paper, pulp and print
- Wood and wood products (other than pulp and paper)
- Textile and leather
- Construction
- Industries, not elsewhere specified

By transport refers to the consumption of electricity by any economic entity to transport goods or persons between points of departure and destination within the national territory.

By households and other consumers: This group consists of energy consumers not classified in Manufacturing, construction and non-fuel mining industries, and consists of the following subgroups:
- Agriculture, Forestry, Fishing
- Commerce and public services
- Households
- Not elsewhere specified

Middle block – Net installed capacity

Net installed capacity refers to the Net maximum electrical capacity, which is the maximum active power that can be supplied continuously, with all plants running, at the point of outlet (i.e., after taking the power supplies for the station auxiliaries and allowing for the losses in those transformers considered integral to the station).

This assumes no restriction of interconnection to the network, but does not include overload capacity that can only be sustained for a short period of time (e.g., internal combustion engines momentarily running above their rated capacity).

The net maximum electricity-generating capacity represents the sum of all individual plants' maximum capacities available to run continuously throughout a prolonged period of operation in a day.

As mentioned for electricity production, a distinction is made between main activity producers and autoproducers. In the *Electricity Profiles*, autoproducer figures are not provided explicitly, but can be easily derived as the difference between total producers and main activity producers.

Bottom block – Combustible fuel input, output and efficiency

As defined in the section about the top block, electricity from **combustible fuels** refers to the production of electricity from the combustion of fuels which are capable of igniting or burning, i.e. reacting with oxygen to produce a significant rise in temperature.

Therefore, for some countries and areas this block may be empty if such processes are not utilized in the country or area in question.

Combustible fuel input is the energy content (in Terajoules) of those fuels which are combusted to generate electricity. This input is broken down by fuel (as applicable) and is calculated from the fuel quantities expressed in their original units (e.g., thousand metric tons) through the application of conversion factors.

Up to 10 fuels (by input contribution in the latest year available) are displayed alphabetically, with the remaining ones, if any, aggregated in the category Others.

Total input is the total aggregate of combustible fuel input.

Total production is the total electricity production from combustible fuels (both main activity and autoproducers), expressed in Terajoules by using the equivalence: 1 GWh = 3.6 Terajoules.

The **estimated efficiency** (of total electricity generation from combustible fuels) is calculated by dividing the Total production by the Total input. It gives an idea of how much energy from the fuels is converted into electricity on average in the electricity and CHP plants in the country or area.

It is important to note that the formula considers all inputs to CHP plants but only the electricity output from these plants and not the heat generated, so the calculated efficiency should be used with caution to compare countries with differing levels of CHP generation (and these levels are not apparent from this publication).

ix

CONCEPTS AND DEFINITIONS

For ease of reference, the table below displays typical efficiencies for electricity and CHP plants for the main types of fuel used as input.

Generation process	Typical plant efficiency
Electricity plants	
Coal	32-40%
Oil	35-40%
Natural gas	45-55%
Biomass	20-30%
CHP plants	
Coal	50-75%
Oil	50-75%
Natural gas	60-90%
Biomass	60-85%

ABBREVIATIONS AND SYMBOLS

The following abbreviations and symbols have been used in the publication:

CHP	Combined Heat and Power
GCV	Gross Calorific Value
GJ	gigajoule (10^9 joules)
GWh	gigawatt-hour (10^6 kWh)
IRES	International Recommendations for Energy Statistics
kcal	kilocalorie
kg	kilogramme
kJ	kilojoule (10^3 joules)
kWh	kilowatt-hour
NCV	Net Calorific Value
t	metric ton
TJ	terajoule (10^{12} joules)
UNSD	United Nations Statistics Division
*	estimate by the United Nations Statistics Division
..	data not applicable or not available

ELECTRICITY PROFILES

Statistics on electricity

Afghanistan

Item	2009	2010	2011	2012	2013	2014
Production, trade and consumption	**Million kilowatt-hours**					
Total main activity and autoproducer	889	892	1005	1022	*1022	*1049
Combustible fuels	114	125	176	144	*218	*154
Hydro	775	767	829	878	*804	*895
Nuclear	..	..	..	..	..	..
Other	..	..	..	..	..	..
Main activity	619	622	719	721	*746	*742
Combustible fuels	109	120	171	139	*213	*149
Hydro	510	502	548	582	*533	*593
Nuclear	..	..	..	..	..	..
Other	..	..	..	..	..	..
Own use in electricity, CHP and heat plants	53	27	286	206	*266	*273
Net production	836	865	719	816	*756	*776
Imports	957	1867	2732	3071	3615	3711
Exports	..	..	..	..	..	..
Losses	*200	*404	404	404	*490	*500
Consumption	*1600	2400	2300	3265	3022	*3767
Energy industries own use	..	..	..	..	..	..
By industry and construction	*300	*500	600	600	*800	*800
By transport	..	..	..	..	..	..
By households and other cons.	*1300	1900	1700	2665	2222	*2967
Net installed capacity	**Thousand kilowatts**					
Total main activity and autoproducer	*489	*489	431	431	*431	*431
Combustible fuels	115	115	115	115	*115	*115
Hydro	*374	*374	316	316	*316	*316
Nuclear	..	..	..	..	..	..
Other	..	..	..	..	..	..
Main activity	*307	*307	249	249	*249	*249
Combustible fuels	60	60	60	60	*60	*60
Hydro	*247	*247	189	189	*189	*189
Nuclear	..	..	..	..	..	..
Other	..	..	..	..	..	..
Combustible fuel input	**Terajoules**					
Hard Coal	*774	774	1290	1290	*1290	*1290
Gas-diesel oil	*860	860	860	860	*860	*869
Fuel oil	*40	*40	*40	*40	*40	*40
Total input	*1674	1674	2190	2190	*2190	*2199
Total production	410	450	634	518	*786	*554
Estimated efficiency (% of production to input)	25	27	29	24	36	25

Statistics on electricity

Albania

Item	2009	2010	2011	2012	2013	2014
Production, trade and consumption	**Million kilowatt-hours**					
Total main activity and autoproducer	5231	7744	4217	4725	6956	4724
Combustible fuels	1	1	59	..	..	..
Hydro	5230	7743	4158	4725	6956	4724
Nuclear	..	..	..	..	..	..
Other	..	..	..	..	..	..
Main activity	5231	7744	4217	4725	6956	4724
Combustible fuels	1	1	59	..	..	..
Hydro	5230	7743	4158	4725	6956	4724
Nuclear	..	..	..	..	..	..
Other	..	..	..	..	..	..
Own use in electricity, CHP and heat plants	1	1	59	0	0	0
Net production	5230	7743	4158	4725	6956	4724
Imports	1885	1985	3262	2538	2323	3250
Exports	486	2934	232	0	0	183
Losses	1250	1121	1076	1119	1953	1119
Consumption	5379	5673	6112	6144	7325	6672
Energy industries own use	5	5	455	384	433	149
By industry and construction	829	1035	1053	1192	1388	1466
By transport	..	..	..	..	..	..
By households and other cons.	4545	4633	4604	4568	5505	5057
Net installed capacity	**Thousand kilowatts**					
Total main activity and autoproducer	1597	1613	1613	1767	1918	1865
Combustible fuels	123	123	123	123	123	123
Hydro	1460	1476	1476	1629	1781	1726
Nuclear	..	..	..	..	..	..
Other	14	14	14	15	14	16
Main activity	1572	1588	1588	1742	1893	1840
Combustible fuels	98	98	98	98	98	98
Hydro	1460	1476	1476	1629	1781	1726
Nuclear	..	..	..	..	..	..
Other	14	14	14	15	14	16
Combustible fuel input	**Terajoules**					
Refinery gas	0	0	0	0	0	0
Other oil products	68	201	429	0	0	0
Total input	68	201	429	0	0	0
Total production	3	3	213	..	..	..
Estimated efficiency (% of production to input)	5	2	50	..	..	..

Statistics on electricity

Algeria

Item	2009	2010	2011	2012	2013	2014
Production, trade and consumption	**Million kilowatt-hours**					
Total main activity and autoproducer	38501	45734	51224	57397	59890	64242
Combustible fuels	38195	45560	50722	56775	59560	63988
Hydro	306	174	502	622	330	254
Nuclear	..	..	..	..	..	..
Other	..	..	..	..	..	..
Main activity	38207	45173	48872	54086	56148	60500
Combustible fuels	37901	44999	48370	53464	55818	60246
Hydro	306	174	502	622	330	254
Nuclear	..	..	..	..	..	..
Other	..	..	..	..	..	..
Own use in electricity, CHP and heat plants	2133	2505	4542	4753	4873	6567
Net production	36368	43229	46682	52644	55017	57675
Imports	330	736	657	936	295	686
Exports	362	803	799	985	384	877
Losses	7859	9091	9899	11068	11023	11003
Consumption	28476	34069	36639	41527	43908	46481
Energy industries own use	565	599	772	750	750	748
By industry and construction	10116	12700	13138	14961	15670	16020
By transport	540	634	696	783	796	867
By households and other cons.	17255	20136	22033	25033	26692	28846
Net installed capacity	**Thousand kilowatts**					
Total main activity and autoproducer	11325	11332	*11545	*13087	*15097	*15957
Combustible fuels	11050	11057	*11270	*12787	*14797	*15657
Hydro	275	275	*275	*300	*300	*300
Nuclear	..	..	..	..	..	..
Other	..	..	..	..	..	..
Main activity	11027	11034	*10391	*10470	*12078	*12766
Combustible fuels	10752	10759	*10116	*10170	*11778	*12466
Hydro	275	275	*275	*300	*300	*300
Nuclear	..	..	..	..	..	..
Other	..	..	..	..	..	..
Combustible fuel input	**Terajoules**					
Gas-diesel oil	7697	13029	34658	47214	52202	15695
Natural gas	477300	477739	515864	558234	536620	623247
Total input	484997	490768	550522	605448	588822	638942
Total production	137502	164016	182599	204390	214416	230357
Estimated efficiency (% of production to input)	28	33	33	34	36	36

Statistics on electricity

American Samoa

Item	2009	2010	2011	2012	2013	2014
Production, trade and consumption	**Million kilowatt-hours**					
Total main activity and autoproducer	178	159	155	157	*156	157
Combustible fuels	178	159	155	156	*155	156
Hydro	..	..	..	..	..	..
Nuclear	..	..	..	..	..	..
Other	..	..	..	1	*1	*1
Main activity	178	159	155	157	*156	157
Combustible fuels	178	159	155	156	*155	156
Hydro	..	..	..	..	..	..
Nuclear	..	..	..	..	..	..
Other	..	..	..	1	*1	*1
Own use in electricity, CHP and heat plants	7	5	7	8	*7	10
Net production	172	154	148	149	*149	147
Imports	..	..	..	..	..	..
Exports	..	..	..	..	..	..
Losses	14	12	12	15	14	13
Consumption	158	142	136	134	136	135
Energy industries own use	..	..	..	..	..	..
By industry and construction	37	21	22	23	23	23
By transport	..	..	..	..	..	..
By households and other cons.	122	121	114	112	112	112
Net installed capacity	**Thousand kilowatts**					
Total main activity and autoproducer	49	45	41	*42	*42	42
Combustible fuels	49	45	41	*41	*41	41
Hydro	..	..	..	..	..	..
Nuclear	..	..	..	..	..	..
Other	..	0	0	1	*1	1
Main activity	49	45	41	*42	*42	42
Combustible fuels	49	45	41	*41	*41	41
Hydro	..	..	..	..	..	..
Nuclear	..	..	..	..	..	..
Other	..	0	0	1	*1	1
Combustible fuel input	**Terajoules**					
Total input	..	..	..	..	..	..
Total production	642	573	558	560	*559	561
Estimated efficiency (% of production to input)	..	..	..	..	..	..

Statistics on electricity

Andorra

Item	2009	2010	2011	2012	2013	2014
Production, trade and consumption	Million kilowatt-hours					
Total main activity and autoproducer	*81	113	91	88	115	127
Combustible fuels	*19	13	12	12	13	13
Hydro	*62	100	79	76	102	113
Nuclear	..	..	..	..	..	..
Other	..	..	..	..	..	..
Main activity	*81	113	91	88	115	127
Combustible fuels	*19	13	12	12	13	13
Hydro	*62	100	79	76	102	113
Nuclear	..	..	..	..	..	..
Other	..	..	..	..	..	..
Own use in electricity, CHP and heat plants	0	0	0	0	0	0
Net production	*81	113	91	88	115	127
Imports	*506	471	466	474	438	415
Exports	..	..	..	..	3	..
Losses	76	78	75	74	71	67
Consumption	493	506	483	488	482	474
Energy industries own use	..	..	..	..	..	..
By industry and construction	11	10	9	9	10	8
By transport	..	..	..	..	..	..
By households and other cons.	482	496	474	480	473	467
Net installed capacity	Thousand kilowatts					
Total main activity and autoproducer	*32	*32	*32	*32	*32	*32
Combustible fuels	*5	*5	*5	*5	*5	*5
Hydro	*27	*27	*27	*27	*27	*27
Nuclear	..	..	..	..	..	..
Other	..	..	..	..	..	..
Main activity	*32	*32	*32	*32	*32	*32
Combustible fuels	*5	*5	*5	*5	*5	*5
Hydro	*27	*27	*27	*27	*27	*27
Nuclear	..	..	..	..	..	..
Other	..	..	..	..	..	..
Combustible fuel input	Terajoules					
Municipal waste	*224	*218	*211	*211	*232	*241
Total input	*224	*218	*211	*211	*232	*241
Total production	*68	46	44	42	46	48
Estimated efficiency (% of production to input)	31	21	21	20	20	20

Statistics on electricity

Angola

Item	2009	2010	2011	2012	2013	2014
Production, trade and consumption	**Million kilowatt-hours**					
Total main activity and autoproducer	4735	5449	5651	6204	8216	9480
Combustible fuels	1641	1746	1644	2432	3449	4439
Hydro	3094	3703	4007	3772	4767	5041
Nuclear	..	..	..	..	..	..
Other	..	..	..	..	..	..
Main activity	4411	5104	5326	5847	7743	8934
Combustible fuels	1317	1401	1319	2075	2976	3893
Hydro	3094	3703	4007	3772	4767	5041
Nuclear	..	..	..	..	..	..
Other	..	..	..	..	..	..
Own use in electricity, CHP and heat plants	237	136	139	153	203	234
Net production	4498	5313	5512	6051	8013	9246
Imports	..	..	..	..	..	..
Exports	..	..	..	..	..	..
Losses	474	627	637	699	926	1068
Consumption	4024	4686	4875	5352	7087	8178
Energy industries own use	..	..	..	..	..	..
By industry and construction	1278	1578	1643	1804	2389	2757
By transport	..	..	..	..	..	..
By households and other cons.	2746	3108	3232	3548	4698	5421
Net installed capacity	**Thousand kilowatts**					
Total main activity and autoproducer	*1155	1439	*1530	*1530	*1530	*2070
Combustible fuels	*657	*657	*770	*770	*770	*770
Hydro	498	782	760	760	*760	*1300
Nuclear	..	..	..	..	..	..
Other	..	..	..	..	..	..
Main activity	*1025	1309	1400	1400	*1400	*1940
Combustible fuels	*527	*527	*640	*640	*640	*640
Hydro	498	782	760	760	*760	*1300
Nuclear	..	..	..	..	..	..
Other	..	..	..	..	..	..
Combustible fuel input	**Terajoules**					
Gas-diesel oil	18662	19866	18705	18576	28294	43000
Fuel oil	10827	11514	10827	5939	2788	3030
Total input	29489	31380	29532	24515	31082	46030
Total production	5908	6286	5918	8755	12416	15980
Estimated efficiency (% of production to input)	20	20	20	36	40	35

Statistics on electricity

Anguilla

Item	2009	2010	2011	2012	2013	2014
Production, trade and consumption	**Million kilowatt-hours**					
Total main activity and autoproducer	91	99	95	90	89	90
Combustible fuels	91	99	95	90	89	90
Hydro	..	..	..	..	..	..
Nuclear	..	..	..	..	..	..
Other	..	..	..	..	..	..
Main activity	91	99	95	90	89	90
Combustible fuels	91	99	95	90	89	90
Hydro	..	..	..	..	..	..
Nuclear	..	..	..	..	..	..
Other	..	..	..	..	..	..
Own use in electricity, CHP and heat plants	1	2	3	3	3	3
Net production	90	97	92	87	86	87
Imports	..	..	..	..	..	..
Exports	..	..	..	..	..	..
Losses	10	9	9	8	9	9
Consumption	*80	*87	*84	*79	*77	*78
Energy industries own use	..	..	..	..	..	..
By industry and construction	..	..	..	..	..	..
By transport	..	..	..	..	..	..
By households and other cons.	*80	*87	*84	*79	*77	*78
Net installed capacity	**Thousand kilowatts**					
Total main activity and autoproducer	28	28	28	28	28	28
Combustible fuels	28	28	28	28	28	28
Hydro	..	..	..	..	..	..
Nuclear	..	..	..	..	..	..
Other	..	..	..	..	..	..
Main activity	28	28	28	28	28	28
Combustible fuels	28	28	28	28	28	28
Hydro	..	..	..	..	..	..
Nuclear	..	..	..	..	..	..
Other	..	..	..	..	..	..
Combustible fuel input	**Terajoules**					
Gas-diesel oil	826	886	856	813	796	821
Total input	826	886	856	813	796	821
Total production	328	355	343	322	319	322
Estimated efficiency (% of production to input)	40	40	40	40	40	39

Statistics on electricity

Antigua and Barbuda

Item	2009	2010	2011	2012	2013	2014
Production, trade and consumption	**Million kilowatt-hours**					
Total main activity and autoproducer	327	*326	*325	*327	*330	*336
Combustible fuels	327	*326	*325	*327	*330	*336
Hydro	..	..	..	..	..	..
Nuclear	..	..	..	..	..	..
Other	..	..	..	..	..	..
Main activity	326	*325	*324	*326	*329	*335
Combustible fuels	326	*325	*324	*326	*329	*335
Hydro	..	..	..	..	..	..
Nuclear	..	..	..	..	..	..
Other	..	..	..	..	..	..
Own use in electricity, CHP and heat plants	7	*7	*7	*7	*7	*7
Net production	320	*319	*318	*320	*323	*329
Imports	..	..	..	..	..	..
Exports	..	..	..	..	..	..
Losses	*85	*84	*84	*85	*85	*87
Consumption	217	*221	*230	*235	*237	*242
Energy industries own use	..	..	..	..	..	..
By industry and construction	6	*7	*8	*8	*8	*9
By transport	..	..	..	..	..	..
By households and other cons.	211	*214	*222	*227	*229	*233
Net installed capacity	**Thousand kilowatts**					
Total main activity and autoproducer	55	55	84	84	*84	*84
Combustible fuels	55	55	84	84	*84	*84
Hydro	..	..	..	..	..	..
Nuclear	..	..	..	..	..	..
Other	..	..	..	..	..	..
Main activity	54	54	83	83	*83	*83
Combustible fuels	54	54	83	83	*83	*83
Hydro	..	..	..	..	..	..
Nuclear	..	..	..	..	..	..
Other	..	..	..	..	..	..
Combustible fuel input	**Terajoules**					
Gas-diesel oil	*2000	*2408	*2365	*2365	*2365	*2365
Fuel oil	*1333	*1374	*1374	*1414	*1414	*1414
Total input	*3333	*3782	*3739	*3779	*3779	*3779
Total production	1179	*1174	*1170	*1177	*1188	*1210
Estimated efficiency (% of production to input)	35	31	31	31	31	32

Statistics on electricity

Argentina

Item	2009	2010	2011	2012	2013	2014
Production, trade and consumption	Million kilowatt-hours					
Total main activity and autoproducer	122348	125594	129892	135207	140863	141586
Combustible fuels	79856	84480	91592	98719	100758	93736
Hydro	34295	33918	31901	29716	33422	41348
Nuclear	8161	7171	6371	6395	6207	5756
Other	36	25	28	377	476	746
Main activity	107892	110269	114840	119978	126147	126883
Combustible fuels	65491	69247	76598	83548	86095	79083
Hydro	34204	33826	31843	29658	33369	41298
Nuclear	8161	7171	6371	6395	6207	5756
Other	36	25	28	377	476	746
Own use in electricity, CHP and heat plants	3420	4076	4304	3919	4303	3900
Net production	118928	121518	125588	131288	136560	137686
Imports	8600	10299	10929	8116	8303	10024
Exports	2445	1701	1262	506	247	166
Losses	17980	16817	18701	18442	19992	20248
Consumption	107096	113308	116551	120569	124624	127300
Energy industries own use	715	551	561	719	497	543
By industry and construction	45520	48331	50458	50535	52859	51881
By transport	664	669	675	616	594	595
By households and other cons.	60197	63757	64857	68699	70674	74281
Net installed capacity	Thousand kilowatts					
Total main activity and autoproducer	32125	32876	33810	35023	39563	*35945
Combustible fuels	21035	21783	22660	23807	28269	*24651
Hydro	10044	10045	10045	10053	10074	10074
Nuclear	1018	1018	1018	1018	1018	1018
Other	28	30	*87	145	202	202
Main activity	28330	29010	29887	31008	35500	*31931
Combustible fuels	17260	17939	18759	19815	24228	*20659
Hydro	10024	10025	10025	10033	10054	10054
Nuclear	1018	1018	1018	1018	1018	1018
Other	28	28	*85	143	200	200
Combustible fuel input	Terajoules					
Hard Coal	13814	24393	30694	35805	26024	36826
Gas-diesel oil	34658	64500	78905	73229	95976	72627
Fuel oil	100030	93647	108151	117968	93243	113160
Refinery gas	5445	4010	1337	2129	2129	3663
Natural gas	616394	586554	626708	688255	624838	720541
Blast furnace gas	3826	4899	5350	5930	5568	5568
Fuelwood	9252	9270	10500	10655	9829	26209
Bagasse	..	..	877	3680	8842	*7000
Vegetal waste	10269	21299	18188	18830	3977	*5500
Biodiesel	..	..	..	1362	1509	1730
Others	4218	5542	5992	3345	2948	967
Total input	797907	814113	886701	961187	874883	993791
Total production	287482	304128	329731	355388	362729	337450
Estimated efficiency (% of production to input)	36	37	37	37	41	34

Statistics on electricity

Armenia

Item	2009	2010	2011	2012	2013	2014
Production, trade and consumption	**Million kilowatt-hours**					
Total main activity and autoproducer	5672	6491	7433	8036	7710	7750
Combustible fuels	1154	1438	2390	3399	3173	3289
Hydro	2020	2556	2489	2311	2173	1992
Nuclear	2494	2490	2548	2322	2360	2465
Other	4	7	6	4	4	4
Main activity	5672	6491	7433	8036	7710	7750
Combustible fuels	1154	1438	2390	3399	3173	3289
Hydro	2020	2556	2489	2311	2173	1992
Nuclear	2494	2490	2548	2322	2360	2465
Other	4	7	6	4	4	4
Own use in electricity, CHP and heat plants	295	279	327	337	329	361
Net production	5377	6212	7106	7699	7381	7389
Imports	295	246	205	98	148	206
Exports	325	1077	1533	1696	1313	1314
Losses	843	730	904	981	949	929
Consumption	4487	4667	5151	5120	5404	5353
Energy industries own use	..	..	..	..	..	..
By industry and construction	1009	1047	1083	1180	1209	1479
By transport	119	119	120	127	124	115
By households and other cons.	3359	3501	3948	3813	4071	3759
Net installed capacity	**Thousand kilowatts**					
Total main activity and autoproducer	3175	3480	3509	4055	4095	4091
Combustible fuels	1662	1906	1906	2394	2394	2390
Hydro	1102	1162	1191	1249	1289	1289
Nuclear	408	408	408	408	408	408
Other	3	4	4	4	4	4
Main activity	3175	3480	3509	4055	4095	4091
Combustible fuels	1662	1906	1906	2394	2394	2390
Hydro	1102	1162	1191	1249	1289	1289
Nuclear	408	408	408	408	408	408
Other	3	4	4	4	4	4
Combustible fuel input	**Terajoules**					
Natural gas	12781	*12532	*20829	*29622	*28760	*28988
Total input	12781	*12532	*20829	*29622	*28760	*28988
Total production	4154	5177	8604	12236	11423	11840
Estimated efficiency (% of production to input)	33	41	41	41	40	41

11

Statistics on electricity

Aruba

Item	2009	2010	2011	2012	2013	2014
Production, trade and consumption	**Million kilowatt-hours**					
Total main activity and autoproducer	924	940	931	915	*928	*917
Combustible fuels	924	834	818	778	*790	*778
Hydro	..	..	..	..	..	..
Nuclear	..	..	..	..	..	..
Other	..	107	113	136	*138	*138
Main activity	924	940	931	915	*928	*917
Combustible fuels	924	834	818	778	*790	*778
Hydro	..	..	..	..	..	..
Nuclear	..	..	..	..	..	..
Other	..	107	113	136	*138	*138
Own use in electricity, CHP and heat plants	0	0	0	0	0	0
Net production	924	940	931	915	*928	*917
Imports	..	..	..	..	..	..
Exports	..	..	..	..	..	..
Losses	150	151	159	148	149	*138
Consumption	*774	*790	*772	*766	*779	*779
Energy industries own use	..	..	..	..	..	..
By industry and construction	..	..	..	..	..	..
By transport	..	..	..	..	..	..
By households and other cons.	*774	*790	*772	*766	*779	*779
Net installed capacity	**Thousand kilowatts**					
Total main activity and autoproducer	236	320	320	320	288	288
Combustible fuels	236	290	290	290	258	258
Hydro	..	..	..	..	..	..
Nuclear	..	..	..	..	..	..
Other	..	30	30	30	30	30
Main activity	236	320	320	320	288	288
Combustible fuels	236	290	290	290	258	258
Hydro	..	..	..	..	..	..
Nuclear	..	..	..	..	..	..
Other	..	30	30	30	30	30
Combustible fuel input	**Terajoules**					
Gas-diesel oil	*1871	*843	*667	*559	*559	*568
Fuel oil	*9454	*8767	*8888	*8363	*7595	*7716
Total input	*11324	*9610	*9555	*8922	*8154	*8284
Total production	3327	3001	2944	2801	*2843	*2802
Estimated efficiency (% of production to input)	29	31	31	31	35	34

Statistics on electricity

Australia

Item	2009	2010	2011	2012	2013	2014
Production, trade and consumption	Million kilowatt-hours					
Total main activity and autoproducer	248754	252698	253958	251165	249720	248299
Combustible fuels	232900	233670	229535	227552	219663	214767
Hydro	11869	13549	16807	14083	18270	18421
Nuclear	..	..	..	..	..	..
Other	3985	5479	7616	9530	11787	15111
Main activity	232681	234165	234144	231498	226356	223944
Combustible fuels	216983	215557	211246	210440	200120	195263
Hydro	11869	13549	16807	14083	18270	18421
Nuclear	..	..	..	..	..	..
Other	3829	5059	6091	6975	7966	10260
Own use in electricity, CHP and heat plants	16612	15278	13081	14351	13848	14831
Net production	232142	237420	240877	236814	235872	233468
Imports	..	..	..	..	..	..
Exports	..	..	..	..	..	..
Losses	14666	16365	16544	14561	13330	11866
Consumption	217475	220929	224241	222149	222335	221601
Energy industries own use	10985	10924	12004	12232	12155	13546
By industry and construction	78708	82103	81502	80016	79645	79483
By transport	4252	3669	3835	4071	4774	4770
By households and other cons.	123530	124233	126900	125830	125761	123802
Net installed capacity	Thousand kilowatts					
Total main activity and autoproducer	57059	60611	62357	64164	64688	66557
Combustible fuels	46628	49571	50044	50377	50171	50704
Hydro	8619	8773	8788	8790	8037	8048
Nuclear	..	..	..	..	..	..
Other	1812	2267	3525	4997	6480	7805
Main activity	52987	55730	56490	57566	56523	57655
Combustible fuels	42661	45089	45571	46211	45261	45806
Hydro	8619	8773	8788	8790	8037	8048
Nuclear	..	..	..	..	..	..
Other	1707	1868	2131	2565	3225	3801
Combustible fuel input	Terajoules					
Hard Coal	496097	785265	735006	717055	696294	648712
Brown Coal	1611072	1205317	1190221	1198113	1024432	955774
Lignite briquettes	1405	1155	1176	1386	1029	861
Biogas	11987	11912	12486	13202	11743	14005
Gas-diesel oil	28853	49450	47128	28079	51987	37195
Fuel oil	5979	5575	5454	5414	7030	6262
Other oil products	2814	3497	3337	2211	3658	2894
Natural gas	428150	454153	468931	514586	517482	530826
Fuelwood	*1957	*1948	*1301	*1315	*1182	*1731
Bagasse	*19637	*16864	*11260	*11385	*10233	*14980
Others	3186	3442	3061	2883	0	0
Total input	2611137	2538578	2479361	2495628	2325070	2213240
Total production	838440	841212	826326	819187	790787	773161
Estimated efficiency (% of production to input)	32	33	33	33	34	35

13

Statistics on electricity

Austria

Item	2009	2010	2011	2012	2013	2014
Production, trade and consumption	**Million kilowatt-hours**					
Total main activity and autoproducer	69088	71128	65813	72617	68277	65421
Combustible fuels	23397	27398	25909	22101	18752	15949
Hydro	43669	41558	37782	47705	45777	44828
Nuclear	..	..	..	..	..	..
Other	2022	2172	2122	2811	3748	4644
Main activity	60603	61649	56359	64045	60178	57730
Combustible fuels	16184	18995	17432	14078	11219	8848
Hydro	42414	40500	36816	47167	45225	44251
Nuclear	..	..	..	..	..	..
Other	2005	2154	2111	2800	3734	4631
Own use in electricity, CHP and heat plants	2641	2989	3116	3080	3417	3083
Net production	66447	68139	62697	69537	64860	62338
Imports	19542	19898	24972	23264	24960	26712
Exports	18762	17567	16777	20455	17689	17437
Losses	3582	3351	3307	3359	3388	3284
Consumption	63645	67119	67585	68987	68743	68329
Energy industries own use	5942	6802	7392	7900	7732	7861
By industry and construction	23882	25835	26357	26499	26568	26565
By transport	3293	3429	3110	3044	3073	3017
By households and other cons.	30528	31053	30726	31544	31370	30886
Net installed capacity	**Thousand kilowatts**					
Total main activity and autoproducer	20790	21187	22556	22917	23591	24025
Combustible fuels	7278	7345	8178	8161	8170	7859
Hydro	12446	12706	12980	13076	13149	13293
Nuclear	..	..	..	..	..	..
Other	1066	1136	1398	1680	2272	2873
Main activity	18675	18876	20182	20568	21212	21768
Combustible fuels	5712	5582	6339	6347	6326	6136
Hydro	11897	12158	12445	12541	12614	12760
Nuclear	..	..	..	..	..	..
Other	1066	1136	1398	1680	2272	2872
Combustible fuel input	**Terajoules**					
Hard Coal	34326	44776	48302	39448	38085	27024
Biogas	5668	5670	5942	6978	6547	10270
Fuel oil	13090	14302	11433	8686	8484	6262
Natural gas	102872	115073	101925	85531	64316	52928
Coke-oven gas	2733	2584	2497	2592	2017	2703
Blast furnace gas	10294	14468	14979	14233	16294	16217
Fuelwood	*40464	*43213	*44856	*48880	*41984	*42709
Municipal waste	11683	12507	13429	13309	13552	15903
Industrial waste	4562	5402	5766	6619	6491	5911
Black liquor	*199	*1728	*4048	*6909	*5099	*5100
Others	2639	2059	1118	1218	1027	1218
Total input	228530	261781	254295	234403	203896	186245
Total production	84229	98633	93272	79564	67507	57416
Estimated efficiency (% of production to input)	37	38	37	34	33	31

Statistics on electricity

Azerbaijan

Item	2009	2010	2011	2012	2013	2014
Production, trade and consumption	**Million kilowatt-hours**					
Total main activity and autoproducer	18868	18710	20294	22988	23354	24728
Combustible fuels	16558	15263	17618	21167	21863	23423
Hydro	2308	3446	2676	1821	1489	1300
Nuclear	..	..	..	..	..	..
Other	2	1	0	0	2	5
Main activity	18599	18450	19993	21358	21556	22706
Combustible fuels	16289	15003	17317	19537	20065	21401
Hydro	2308	3446	2676	1821	1489	1300
Nuclear	..	..	..	..	..	..
Other	2	1	0	0	2	5
Own use in electricity, CHP and heat plants	960	975	947	977	961	998
Net production	17908	17735	19347	22011	22393	23730
Imports	110	100	128	141	127	124
Exports	380	462	805	680	495	489
Losses	4100	3830	3973	3368	3281	3363
Consumption	13436	13419	14567	18023	18661	19914
Energy industries own use	1177	1184	1300	2629	2679	3007
By industry and construction	1904	1758	2130	3027	2996	3161
By transport	521	545	545	523	531	536
By households and other cons.	9834	9932	10592	11844	12455	13210
Net installed capacity	**Thousand kilowatts**					
Total main activity and autoproducer	6388	6396	6350	6420	7353	7355
Combustible fuels	5401	5401	5352	5397	6264	6270
Hydro	987	995	998	1023	1084	1078
Nuclear	..	..	..	..	..	..
Other	..	0	0	0	5	7
Main activity	6288	6296	6250	6294	7117	7113
Combustible fuels	5301	5301	5252	5271	6028	6028
Hydro	987	995	998	1023	1084	1078
Nuclear	..	..	..	..	..	..
Other	..	0	0	0	5	7
Combustible fuel input	**Terajoules**					
Gas-diesel oil	129	129	129	215	258	215
Fuel oil	5212	81	3313	3959	0	162
Natural gas	175521	161896	180185	203578	206854	214193
Municipal waste	..	..	..	..	2104	2628
Total input	180862	162106	183627	207752	209216	217198
Total production	59609	54947	63425	76201	78707	84323
Estimated efficiency (% of production to input)	33	34	35	37	38	39

Statistics on electricity

Bahamas

Item	2009	2010	2011	2012	2013	2014
Production, trade and consumption	**Million kilowatt-hours**					
Total main activity and autoproducer	2140	1805	1754	1750	1885	1853
Combustible fuels	2140	1805	1754	1750	1885	1853
Hydro	..	..	..	..	..	..
Nuclear	..	..	..	..	..	..
Other	..	..	..	..	..	..
Main activity	2068	1752	1703	1699	1811	1779
Combustible fuels	2068	1752	1703	1699	1811	1779
Hydro	..	..	..	..	..	..
Nuclear	..	..	..	..	..	..
Other	..	..	..	..	..	..
Own use in electricity, CHP and heat plants	72	0	0	0	74	74
Net production	2068	1805	1754	1750	1811	1779
Imports	..	..	..	..	..	..
Exports	..	..	..	..	..	..
Losses	..	222	216	215	..	..
Consumption	1787	1567	1523	1520	1767	1808
Energy industries own use	..	..	..	..	..	..
By industry and construction	987	113	110	110	1048	1004
By transport	..	..	..	..	..	..
By households and other cons.	800	1454	1413	1410	719	804
Net installed capacity	**Thousand kilowatts**					
Total main activity and autoproducer	591	591	591	591	591	576
Combustible fuels	591	591	591	591	591	576
Hydro	..	..	..	..	..	..
Nuclear	..	..	..	..	..	..
Other	..	..	..	..	..	..
Main activity	536	536	536	536	536	537
Combustible fuels	536	536	536	536	536	537
Hydro	..	..	..	..	..	..
Nuclear	..	..	..	..	..	..
Other	..	..	..	..	..	..
Combustible fuel input	**Terajoules**					
Gas-diesel oil	*15480	13717	13889	11008	*12814	*15308
Liquefied petroleum gas	..	33	19	47	..	..
Total input	*15480	13750	13908	11055	*12814	*15308
Total production	7704	6497	6313	6298	6786	6671
Estimated efficiency (% of production to input)	50	47	45	57	53	44

Bahrain

Item	2009	2010	2011	2012	2013	2014
Production, trade and consumption	Million kilowatt-hours					
Total main activity and autoproducer	22555	23441	24338	24770	25909	27246
Combustible fuels	22555	23441	24338	24770	25909	27246
Hydro	..	..	..	..	..	..
Nuclear	..	..	..	..	..	..
Other	..	..	..	..	..	..
Main activity	12056	13230	13706	14260	14743	16257
Combustible fuels	12056	13230	13706	14260	14743	16257
Hydro	..	..	..	..	..	..
Nuclear	..	..	..	..	..	..
Other	..	..	..	..	..	..
Own use in electricity, CHP and heat plants	175	193	120	117	62	-4
Net production	22380	23248	24218	24653	25847	27250
Imports	168	192	227	35	70	240
Exports	0	19	107	190	53	237
Losses	2038	1614	1563	1417	1351	1073
Consumption	20510	22193	22775	23315	24513	26186
Energy industries own use	..	..	..	..	..	..
By industry and construction	11568	11458	12333	12410	13180	13633
By transport	..	..	..	..	..	..
By households and other cons.	8942	10735	10442	10905	11333	12553
Net installed capacity	Thousand kilowatts					
Total main activity and autoproducer	*5868	*5868	*5868	*5868	*5868	*6050
Combustible fuels	*5868	*5868	*5868	*5868	*5868	*6050
Hydro	..	..	..	..	..	..
Nuclear	..	..	..	..	..	..
Other	..	..	..	..	..	..
Main activity	*2868	*2868	*2868	*2868	*2868	*3000
Combustible fuels	*2868	*2868	*2868	*2868	*2868	*3000
Hydro	..	..	..	..	..	..
Nuclear	..	..	..	..	..	..
Other	..	..	..	..	..	..
Combustible fuel input	Terajoules					
Natural gas	352235	355879	363675	372448	389899	406937
Total input	352235	355879	363675	372448	389899	406937
Total production	81198	84388	87617	89172	93272	98086
Estimated efficiency (% of production to input)	23	24	24	24	24	24

Statistics on electricity

Bangladesh

Item	2009	2010	2011	2012	2013	2014
Production, trade and consumption	**Million kilowatt-hours**					
Total main activity and autoproducer	37177	40790	44153	48562	53043	55845
Combustible fuels	36760	40061	43281	47785	52000	55108
Hydro	417	729	872	777	894	588
Nuclear	..	..	..	..	..	..
Other	..	..	..	..	149	149
Main activity	27137	28754	31202	35354	38766	40427
Combustible fuels	26720	28025	30330	34577	37723	39690
Hydro	417	729	872	777	894	588
Nuclear	..	..	..	..	..	..
Other	..	..	..	..	149	149
Own use in electricity, CHP and heat plants	2231	2447	2649	2914	3110	3351
Net production	34946	38343	41504	45648	49933	52494
Imports	..	..	..	..	..	..
Exports	..	..	..	..	..	..
Losses	4191	4300	4500	5757	6991	6367
Consumption	30793	34392	36971	40088	42717	48918
Energy industries own use	..	..	..	..	..	..
By industry and construction	17568	19552	21202	22163	24000	26681
By transport	..	..	..	..	..	..
By households and other cons.	13225	14840	15769	17925	18717	22237
Net installed capacity	**Thousand kilowatts**					
Total main activity and autoproducer	5719	5827	6639	8100	8562	11557
Combustible fuels	5489	5597	6409	7870	8222	11302
Hydro	230	230	230	230	315	230
Nuclear	..	..	..	..	..	..
Other	..	..	..	..	*25	*25
Main activity	3812	3916	4728	5725	6340	9200
Combustible fuels	3582	3686	4498	5495	6000	8945
Hydro	230	230	230	230	315	230
Nuclear	..	..	..	..	..	..
Other	..	..	..	..	*25	*25
Combustible fuel input	**Terajoules**					
Hard Coal	9856	10024	8580	9396	12388	11300
Gas-diesel oil	6923	7697	9718	9718	8901	15781
Fuel oil	6504	7676	21533	34986	42582	48036
Natural gas	382771	435111	434148	470590	508333	530247
Total input	406055	460508	473979	524690	572204	605364
Total production	132336	144220	155812	172026	187200	198389
Estimated efficiency (% of production to input)	33	31	33	33	33	33

Statistics on electricity

Barbados

Item	2009	2010	2011	2012	2013	2014
Production, trade and consumption	**Million kilowatt-hours**					
Total main activity and autoproducer	1068	1028	1049	1129	1016	1010
Combustible fuels	1068	1028	1049	1129	1016	1010
Hydro	..	..	..	..	..	..
Nuclear	..	..	..	..	..	..
Other	..	..	..	..	..	..
Main activity	1068	1028	1044	1125	1013	1007
Combustible fuels	1068	1028	1044	1125	1013	1007
Hydro	..	..	..	..	..	..
Nuclear	..	..	..	..	..	..
Other	..	..	..	..	..	..
Own use in electricity, CHP and heat plants	72	19	41	37	58	58
Net production	997	1009	1008	1091	958	951
Imports	..	..	..	..	..	..
Exports	..	..	..	..	..	..
Losses	45	51	68	43	43	48
Consumption	951	958	940	1048	915	*903
Energy industries own use	..	..	..	..	..	..
By industry and construction	63	63	86	76	73	79
By transport	..	..	..	..	..	..
By households and other cons.	889	895	854	972	842	*825
Net installed capacity	**Thousand kilowatts**					
Total main activity and autoproducer	215	257	242	241	241	249
Combustible fuels	215	257	242	241	241	249
Hydro	..	..	..	..	..	..
Nuclear	..	..	..	..	..	..
Other	..	..	..	..	..	..
Main activity	215	257	239	239	239	239
Combustible fuels	215	257	239	239	239	239
Hydro	..	..	..	..	..	..
Nuclear	..	..	..	..	..	..
Other	..	..	..	..	..	..
Combustible fuel input	**Terajoules**					
Gas-diesel oil	1707	1846	2034	1777	1689	773
Fuel oil	9685	8008	8538	8682	8632	7413
Other kerosene	335	329	363	271	*336	*282
Natural gas	82	108	115	152	255	0
Bagasse	0	0	43	38	29	23
Total input	11808	10292	11092	10920	10941	8490
Total production	3846	3700	3776	4063	3657	3634
Estimated efficiency (% of production to input)	33	36	34	37	33	43

19

Statistics on electricity

Belarus

Item	2009	2010	2011	2012	2013	2014
Production, trade and consumption	**Million kilowatt-hours**					
Total main activity and autoproducer	30376	34895	32200	30799	31507	34735
Combustible fuels	30331	34849	32157	30723	31361	34602
Hydro	44	45	42	70	138	121
Nuclear	..	..	..	..	..	..
Other	1	1	1	6	8	12
Main activity	28640	32455	29604	27749	28531	31663
Combustible fuels	28610	32412	29566	27681	28394	31541
Hydro	30	43	38	64	131	114
Nuclear	..	..	..	..	..	..
Other	..	..	..	4	6	8
Own use in electricity, CHP and heat plants	2130	2264	2220	2203	2189	2217
Net production	28246	32631	29980	28596	29318	32518
Imports	8404	7767	9289	10398	9382	7806
Exports	3933	5067	3704	2797	3012	4488
Losses	3487	3774	3412	3406	3342	3187
Consumption	29230	31557	32153	32791	32346	32649
Energy industries own use	1539	2175	2257	2413	2411	2429
By industry and construction	12332	13204	13603	13353	12719	12899
By transport	1529	1613	1468	1402	1304	1280
By households and other cons.	13830	14565	14825	15623	15912	16041
Net installed capacity	**Thousand kilowatts**					
Total main activity and autoproducer	*8106	8275	8375	8975	9181	10037
Combustible fuels	*8089	8259	8356	8939	9144	9999
Hydro	*15	15	15	33	33	33
Nuclear	..	..	..	..	..	..
Other	2	2	3	3	*3	*4
Main activity	*7729	7822	7906	8393	8473	9327
Combustible fuels	*7719	7812	7895	8365	8445	9298
Hydro	*10	9	9	26	26	26
Nuclear	..	..	..	..	..	..
Other	0	0	2	2	2	2
Combustible fuel input	**Terajoules**					
Peat	885	854	875	885	1081	1291
Biogas	61	78	116	145	337	363
Gas-diesel oil	0	0	0	0	0	0
Fuel oil	76800	8363	3111	8888	1293	2868
Refinery gas	1040	1634	1832	2277	2030	2228
Natural gas	345512	482698	448015	426772	441450	451493
Fuelwood	2371	2820	3088	3180	3365	3523
Vegetal waste	0	0	0	0	0	0
Industrial waste	528	604	726	674	686	880
Peat products	39	254	224	342	371	332
Others	0	0	0	0	0	0
Total input	427236	497305	457987	443163	450612	462977
Total production	109192	125456	115765	110603	112900	124567
Estimated efficiency (% of production to input)	26	25	25	25	25	27

Statistics on electricity

Belgium

Item	2009	2010	2011	2012	2013	2014
Production, trade and consumption	Million kilowatt-hours					
Total main activity and autoproducer	91225	95189	90241	82923	83526	72688
Combustible fuels	40999	43470	36969	35843	32467	29582
Hydro	1757	1668	1423	1659	1723	1507
Nuclear	47222	47944	48234	40295	42644	33703
Other	1247	2107	3615	5126	6692	7896
Main activity	87401	89619	83367	74427	74402	62899
Combustible fuels	37425	38722	31408	29740	26357	23094
Hydro	1757	1668	1423	1659	1723	1507
Nuclear	47222	47944	48234	40295	42644	33703
Other	997	1285	2302	2733	3678	4595
Own use in electricity, CHP and heat plants	3702	3703	3582	3140	3293	2723
Net production	87523	91486	86659	79783	80233	69965
Imports	9486	12395	13189	16848	17243	21791
Exports	11321	11844	10652	6912	7603	4188
Losses	4065	4283	4154	4131	4006	3879
Consumption	81544	87704	85523	86137	84924	83622
Energy industries own use	4289	4393	4931	4424	3180	3061
By industry and construction	32676	38137	37261	37490	37196	37752
By transport	1762	1736	1631	1581	1686	1593
By households and other cons.	42817	43438	41700	42642	42862	41216
Net installed capacity	Thousand kilowatts					
Total main activity and autoproducer	17785	18690	20098	20773	20984	20918
Combustible fuels	9472	9522	10285	9468	8910	8605
Hydro	1417	1425	1426	1427	1429	1429
Nuclear	5902	5927	5927	5927	5927	5927
Other	994	1816	2460	3951	4718	4957
Main activity	16321	16664	17553	16904	16708	16494
Combustible fuels	8391	8397	9128	8182	7559	7209
Hydro	1417	1425	1426	1427	1429	1429
Nuclear	5902	5927	5927	5927	5927	5927
Other	611	915	1072	1368	1793	1929
Combustible fuel input	Terajoules					
Hard Coal	45177	40799	34444	32825	28423	21770
Biogas	3783	4126	3666	4254	4593	5168
Refinery gas	..	1188	1436	2525	446	1139
Natural gas	231547	246135	187904	186532	163530	148164
Blast furnace gas	9470	14589	15628	16176	16560	16739
Fuelwood	*23575	*26137	*29689	*34228	*31337	*23516
Municipal waste	26189	30872	37361	28588	28059	28262
Industrial waste	8886	8110	7478	8651	6595	6121
Other recovered gases	..	610	988	1245	1238	1479
Other liquid biofuels	2165	1507	1178	877	658	411
Others	3066	3139	2321	1352	1112	777
Total input	353858	377212	322092	317253	282549	253545
Total production	147596	156492	133088	129035	116881	106495
Estimated efficiency (% of production to input)	42	41	41	41	41	42

Statistics on electricity

Belize

Item	2009	2010	2011	2012	2013	2014
Production, trade and consumption	**Million kilowatt-hours**					
Total main activity and autoproducer	*237	*415	*388	*424	*371	*404
Combustible fuels	*49	152	*144	*215	*164	*199
Hydro	*188	*263	*244	*208	*207	*205
Nuclear	..	..	..	..	..	..
Other	..	..	..	..	..	..
Main activity	*237	*323	*322	*352	*308	*335
Combustible fuels	*49	*60	*78	*143	*101	*130
Hydro	*188	*263	*244	*208	*207	*205
Nuclear	..	..	..	..	..	..
Other	..	..	..	..	..	..
Own use in electricity, CHP and heat plants	*-15	*15	*13	*14	*14	*14
Net production	*251	*400	*375	*410	*357	*390
Imports	216	160	171	172	234	233
Exports	..	..	..	..	..	..
Losses	55	57	51	56	66	65
Consumption	412	503	495	526	536	558
Energy industries own use	..	..	..	..	..	..
By industry and construction	28	112	117	123	130	137
By transport	..	..	..	..	..	..
By households and other cons.	384	391	377	403	406	422
Net installed capacity	**Thousand kilowatts**					
Total main activity and autoproducer	112	144	144	144	143	154
Combustible fuels	66	91	91	91	88	99
Hydro	46	53	53	53	55	55
Nuclear	..	..	..	..	..	..
Other	..	..	..	..	..	..
Main activity	112	118	118	118	120	120
Combustible fuels	66	65	65	65	66	66
Hydro	46	53	53	53	55	55
Nuclear	..	..	..	..	..	..
Other	..	..	..	..	..	..
Combustible fuel input	**Terajoules**					
Crude oil	59	18	14	17	17	15
Gas-diesel oil	229	90	85	86	207	76
Fuel oil	*145	147	*147	*162	*174	*174
Natural gas	..	121	42	56	68	40
Bagasse	..	*1676	*1946	*2193	*2139	*2070
Total input	433	*2051	*2235	*2514	*2605	*2374
Total production	*176	547	*518	*775	*590	*717
Estimated efficiency (% of production to input)	41	27	23	31	23	30

Statistics on electricity

Benin

Item	2009	2010	2011	2012	2013	2014
Production, trade and consumption	**Million kilowatt-hours**					
Total main activity and autoproducer	128	150	155	162	173	184
Combustible fuels	128	150	155	162	173	184
Hydro	..	..	..	..	..	..
Nuclear	..	..	..	..	..	..
Other	..	..	..	..	..	..
Main activity	112	123	127	133	142	151
Combustible fuels	112	123	127	133	142	151
Hydro	..	..	..	..	..	..
Nuclear	..	..	..	..	..	..
Other	..	..	..	..	..	..
Own use in electricity, CHP and heat plants	0	0	0	0	0	0
Net production	128	150	155	162	173	184
Imports	866	935	924	967	1033	1101
Exports	..	..	..	..	..	..
Losses	208	207	213	223	238	254
Consumption	786	842	866	906	968	1031
Energy industries own use	..	..	..	..	..	..
By industry and construction	128	137	140	147	157	167
By transport	..	..	..	..	..	..
By households and other cons.	658	705	726	759	811	864
Net installed capacity	**Thousand kilowatts**					
Total main activity and autoproducer	145	99	100	62	208	*208
Combustible fuels	144	98	99	61	207	*207
Hydro	*1	*1	*1	*1	*1	*1
Nuclear	..	..	..	..	..	..
Other	..	..	..	..	..	..
Main activity	137	91	92	54	200	*200
Combustible fuels	136	90	91	53	199	*199
Hydro	*1	*1	*1	*1	*1	*1
Nuclear	..	..	..	..	..	..
Other	..	..	..	..	..	..
Combustible fuel input	**Terajoules**					
Gas-diesel oil	1247	1462	1505	1548	1634	1720
Fuelwood	*30	*22	*23	*24	*25	*26
Total input	1277	1484	1528	1572	1659	1746
Total production	461	540	558	583	623	662
Estimated efficiency (% of production to input)	36	36	37	37	38	38

Statistics on electricity

Bermuda

Item	2009	2010	2011	2012	2013	2014
Production, trade and consumption	**Million kilowatt-hours**					
Total main activity and autoproducer	*746	742	*726	*692	*673	*710
Combustible fuels	*746	742	*726	*692	*673	*710
Hydro	..	..	..	..	..	..
Nuclear	..	..	..	..	..	..
Other	..	..	..	..	..	..
Main activity	*733	730	*714	*680	*661	*698
Combustible fuels	*733	730	*714	*680	*661	*698
Hydro	..	..	..	..	..	..
Nuclear	..	..	..	..	..	..
Other	..	..	..	..	..	..
Own use in electricity, CHP and heat plants	*45	50	*48	*41	*41	*87
Net production	*701	692	*678	*651	*632	*623
Imports	..	..	..	..	..	..
Exports	..	..	..	..	..	..
Losses	*50	*49	*48	*46	*46	*46
Consumption	657	651	636	606	586	577
Energy industries own use	..	..	..	..	..	..
By industry and construction	..	..	..	..	..	..
By transport	..	..	..	..	..	..
By households and other cons.	657	651	636	606	586	577
Net installed capacity	**Thousand kilowatts**					
Total main activity and autoproducer	*169	*169	169	171	*171	*171
Combustible fuels	*169	*169	169	171	*171	*171
Hydro	..	..	..	..	..	..
Nuclear	..	..	..	..	..	..
Other	..	..	..	..	..	..
Main activity	*165	*165	165	167	*167	*167
Combustible fuels	*165	*165	165	167	*167	*167
Hydro	..	..	..	..	..	..
Nuclear	..	..	..	..	..	..
Other	..	..	..	..	..	..
Combustible fuel input	**Terajoules**					
Gas-diesel oil	*993	989	950	*907	*820	*1511
Fuel oil	*4121	*4305	*4363	*4977	*4977	*3344
Municipal waste	*612	*584	*560	*565	*565	*605
Total input	*5726	*5878	*5874	*6450	*6363	*5460
Total production	*2686	2673	*2613	*2491	*2423	*2556
Estimated efficiency (% of production to input)	47	45	44	39	38	47

24

Statistics on electricity

Bhutan

Item	2009	2010	2011	2012	2013	2014
Production, trade and consumption	Million kilowatt-hours					
Total main activity and autoproducer	6998	7328	7068	6827	7640	7004
Combustible fuels	0	0	0	0	1	1
Hydro	6998	7328	7067	6827	7640	7003
Nuclear	..	..	..	..	..	..
Other	..	..	..	..	..	..
Main activity	6998	7328	7068	6827	7640	7004
Combustible fuels	0	0	0	0	1	1
Hydro	6998	7328	7067	6827	7640	7003
Nuclear	..	..	..	..	..	..
Other	..	..	..	..	..	..
Own use in electricity, CHP and heat plants	223	426	106	120	176	114
Net production	6775	6902	6962	6707	7464	6890
Imports	20	20	20	37	108	187
Exports	5352	5268	5268	4890	5648	4992
Losses	15	19	23	23	21	24
Consumption	1498	1573	1638	1770	1841	2004
Energy industries own use	..	..	..	..	..	..
By industry and construction	1110	1322	1365	1462	1519	1661
By transport	..	..	..	..	..	..
By households and other cons.	388	251	273	307	322	343
Net installed capacity	Thousand kilowatts					
Total main activity and autoproducer	1505	1505	1505	1505	1506	1632
Combustible fuels	17	17	17	17	17	17
Hydro	1488	1488	1488	1488	1488	1614
Nuclear	..	..	..	..	..	..
Other	0	0	0	0	0	0
Main activity	1505	1505	1505	1505	1506	1632
Combustible fuels	17	17	17	17	17	17
Hydro	1488	1488	1488	1488	1488	1614
Nuclear	..	..	..	..	..	..
Other	0	0	0	0	0	0
Combustible fuel input	Terajoules					
Gas-diesel oil	*1	*3	*3	*2	*6	*5
Total input	*1	*3	*3	*2	*6	*5
Total production	0	1	1	1	2	2
Estimated efficiency (% of production to input)	33	33	33	33	33	33

Statistics on electricity

Bolivia (Plur. State of)

Item	2009	2010	2011	2012	2013	2014
Production, trade and consumption	**Million kilowatt-hours**					
Total main activity and autoproducer	6121	6777	7220	7661	8065	8755
Combustible fuels	3823	4592	4870	5306	5526	6490
Hydro	2296	2182	2347	2352	2535	2251
Nuclear	..	..	..	..	..	..
Other	2	3	3	3	4	14
Main activity	6000	6658	7098	7538	8006	8563
Combustible fuels	3704	4476	4751	5186	5470	6304
Hydro	2296	2182	2347	2352	2535	2251
Nuclear	..	..	..	..	..	..
Other	..	..	..	..	1	8
Own use in electricity, CHP and heat plants	103	123	140	159	191	192
Net production	6018	6654	7080	7502	7874	8563
Imports	..	..	..	..	..	..
Exports	..	..	..	..	..	..
Losses	718	783	796	869	728	805
Consumption	5717	6112	6329	6613	6962	7392
Energy industries own use	..	..	..	..	..	..
By industry and construction	1756	1717	1734	1790	1882	2016
By transport	..	..	..	..	..	..
By households and other cons.	3961	4395	4595	4823	5080	5376
Net installed capacity	**Thousand kilowatts**					
Total main activity and autoproducer	1531	1647	1684	1882	*2114	*1606
Combustible fuels	1041	1157	1188	1387	*1618	*1143
Hydro	488	488	494	494	*494	*455
Nuclear	..	..	..	..	..	..
Other	*2	*2	*2	*2	*2	8
Main activity	1403	1508	1563	1746	*1959	*1487
Combustible fuels	942	1047	1096	1279	*1492	*1053
Hydro	461	462	467	467	*467	*431
Nuclear	..	..	..	..	..	..
Other	..	..	..	..	..	3
Combustible fuel input	**Terajoules**					
Gas-diesel oil	1333	1677	1462	1978	1892	2236
Natural gas	45959	55467	60161	62116	60107	68418
Vegetal waste	*2846	*2776	*2869	*2880	*1339	*4470
Total input	50138	59920	64492	66974	63338	75124
Total production	13763	16531	17532	19102	19894	23364
Estimated efficiency (% of production to input)	27	28	27	29	31	31

Bonaire, St Eustatius, Saba

Item	2009	2010	2011	2012	2013	2014
Production, trade and consumption				**Million kilowatt-hours**		
Total main activity and autoproducer	..	..	..	*119	*122	*124
Combustible fuels	..	..	..	*119	*122	*122
Hydro	..	..	..	..	..	..
Nuclear	..	..	..	..	..	..
Other	..	..	..	..	..	2
Main activity	..	..	..	*119	*122	*124
Combustible fuels	..	..	..	*119	*122	*122
Hydro	..	..	..	..	..	..
Nuclear	..	..	..	..	..	..
Other	..	..	..	..	..	2
Own use in electricity, CHP and heat plants	..	..	..	*11	*12	*11
Net production	..	..	..	*107	*110	*113
Imports	..	..	..	..	..	..
Exports	..	..	..	..	..	..
Losses	..	..	..	*19	*20	*21
Consumption	..	..	..	*88	*91	*92
Energy industries own use	..	..	..	..	..	..
By industry and construction	..	..	..	*48	*50	*50
By transport	..	..	..	..	..	..
By households and other cons.	..	..	..	*40	*41	*42
Net installed capacity				**Thousand kilowatts**		
Total main activity and autoproducer	..	..	..	*43	*43	*43
Combustible fuels	..	..	..	*14	*14	14
Hydro	..	..	..	..	..	..
Nuclear	..	..	..	..	..	..
Other	..	..	..	*29	*29	*29
Main activity	..	..	..	*43	*43	*43
Combustible fuels	..	..	..	*14	*14	14
Hydro	..	..	..	..	..	..
Nuclear	..	..	..	..	..	..
Other	..	..	..	*29	*29	*29
Combustible fuel input				**Terajoules**		
Fuel oil	..	..	..	*1103	*1164	*1164
Total input	..	..	..	*1103	*1164	*1164
Total production	..	..	..	*427	*439	*439
Estimated efficiency (% of production to input)	..	..	..	39	38	38

Statistics on electricity

Bosnia and Herzegovina

Item	2009	2010	2011	2012	2013	2014
Production, trade and consumption	**Million kilowatt-hours**					
Total main activity and autoproducer	15668	17124	15280	14082	17451	16160
Combustible fuels	9429	9098	10893	9867	10215	10225
Hydro	6239	8026	4387	4215	7236	5935
Nuclear	..	..	..	..	..	..
Other	..	..	..	..	..	..
Main activity	15348	16722	14955	13739	17082	15757
Combustible fuels	9109	8696	10568	9524	9846	9822
Hydro	6239	8026	4387	4215	7236	5935
Nuclear	..	..	..	..	..	..
Other	..	..	..	..	..	..
Own use in electricity, CHP and heat plants	967	966	1053	1047	998	988
Net production	14701	16158	14227	13035	16453	15172
Imports	2887	3076	4171	4481	3167	3162
Exports	5877	6905	5660	4525	6862	5998
Losses	1875	1602	1549	1496	1448	1322
Consumption	9836	10727	11189	11495	11310	11014
Energy industries own use	373	380	401	398	377	427
By industry and construction	3132	3819	4215	4383	4149	3910
By transport	98	136	139	107	84	80
By households and other cons.	6233	6392	6434	6607	6700	6597
Net installed capacity	**Thousand kilowatts**					
Total main activity and autoproducer	*4304	*4304	*4304	*4304	*4304	*4304
Combustible fuels	*2187	*2187	*2187	*2187	*2187	*2187
Hydro	*2117	*2117	*2117	*2117	*2117	*2117
Nuclear	..	..	..	..	..	..
Other	..	..	..	..	..	..
Main activity	*4074	*4074	*4074	*4074	*4074	*4074
Combustible fuels	*1957	*1957	*1957	*1957	*1957	*1957
Hydro	*2117	*2117	*2117	*2117	*2117	*2117
Nuclear	..	..	..	..	..	..
Other	..	..	..	..	..	..
Combustible fuel input	**Terajoules**					
Brown Coal	133127	130495	155858	145338	139422	142903
Fuel oil	364	485	364	283	485	485
Natural gas	815	680	681	719	530	416
Total input	134306	131660	156902	146340	140437	143803
Total production	33944	32753	39215	35521	36774	36810
Estimated efficiency (% of production to input)	25	25	25	24	26	26

Statistics on electricity

Botswana

Item	2009	2010	2011	2012	2013	2014
Production, trade and consumption	Million kilowatt-hours					
Total main activity and autoproducer	621	532	437	251	761	2365
Combustible fuels	621	532	437	250	760	2363
Hydro	..	..	..	..	..	..
Nuclear	..	..	..	..	..	..
Other	..	..	..	1	1	*2
Main activity	621	532	437	251	761	2365
Combustible fuels	621	532	437	250	760	2363
Hydro	..	..	..	..	..	..
Nuclear	..	..	..	..	..	..
Other	..	..	..	1	1	*2
Own use in electricity, CHP and heat plants	71	75	66	31	92	292
Net production	550	457	371	220	669	2073
Imports	2749	2985	3180	3371	2981	1684
Exports	..	..	..	..	..	..
Losses	381	333	434	393	340	255
Consumption	2917	3109	3118	3198	3310	3449
Energy industries own use	..	..	..	..	..	..
By industry and construction	1123	1141	1117	1086	1128	1463
By transport	..	..	..	..	..	..
By households and other cons.	1794	1967	2001	2112	2183	1986
Net installed capacity	Thousand kilowatts					
Total main activity and autoproducer	152	152	152	153	753	621
Combustible fuels	152	152	152	152	752	620
Hydro	..	..	..	..	..	..
Nuclear	..	..	..	..	..	..
Other	..	0	..	1	1	1
Main activity	132	132	132	133	733	601
Combustible fuels	132	132	132	132	732	600
Hydro	..	..	..	..	..	..
Nuclear	..	..	..	..	..	..
Other	..	0	..	1	1	1
Combustible fuel input	Terajoules					
Hard Coal	9934	8495	6985	3988	12152	38534
Gas-diesel oil	0	0	602	1032	1591	1419
Total input	9934	8495	7587	5020	13743	39953
Total production	2235	1916	1574	898	2736	8507
Estimated efficiency (% of production to input)	22	23	21	18	20	21

29

Statistics on electricity

Brazil

Item	2009	2010	2011	2012	2013	2014
Production, trade and consumption	Million kilowatt-hours					
Total main activity and autoproducer	466158	515798	531759	552498	570838	590541
Combustible fuels	60975	95809	85062	116068	157810	189497
Hydro	390988	403289	428333	415342	390992	373439
Nuclear	12957	14523	15659	16038	15450	15378
Other	1238	2177	2705	5050	6586	12227
Main activity	409150	442803	454726	474469	484675	496510
Combustible fuels	23285	43504	30741	58502	93705	117565
Hydro	371670	382599	405621	394879	368939	351351
Nuclear	12957	14523	15659	16038	15450	15378
Other	1238	2177	2705	5050	6581	12216
Own use in electricity, CHP and heat plants	4895	5685	5075	6129	7666	8687
Net production	461263	510113	526684	546369	563172	581854
Imports	40746	35906	38430	40722	40334	33778
Exports	1080	1257	2544	467	0	3
Losses	79795	85747	87524	94355	94995	91759
Consumption	421135	459015	475045	492269	508510	523871
Energy industries own use	13254	21152	18297	20221	22056	22472
By industry and construction	186741	203351	209390	209622	210159	207046
By transport	1591	1662	1700	1785	1884	1979
By households and other cons.	219549	232850	245658	260641	274411	292374
Net installed capacity	Thousand kilowatts					
Total main activity and autoproducer	106573	112400	117134	120973	126743	133913
Combustible fuels	25353	29154	31243	32786	36528	37827
Hydro	78611	80637	82459	84294	86018	89193
Nuclear	2007	2007	2007	2007	1990	1990
Other	602	602	1425	1886	2207	4903
Main activity	94299	96867	99359	102155	106831	113012
Combustible fuels	16480	17013	17905	18448	21426	21800
Hydro	75210	77245	78022	79814	81213	84330
Nuclear	2007	2007	2007	2007	1990	1990
Other	602	602	1425	1886	2202	4892
Combustible fuel input	Terajoules					
Hard Coal	107066	130745	127206	168334	236287	295936
Gas-diesel oil	74949	95331	91934	116874	116788	145039
Fuel oil	45652	45167	29411	54015	91748	145400
Refinery gas	12548	19597	13463	17194	30210	22459
Other oil products	6030	5186	8241	10693	8643	9447
Natural gas	115327	307963	230183	398537	611516	739443
Coke-oven gas	14482	13318	13171	11595	11153	14839
Fuelwood	8295	11592	10752	11391	12570	14141
Bagasse	97407	147963	144372	160527	190829	207176
Vegetal waste	95938	94463	128035	143236	103244	194488
Others	413	1121	0	0	0	0
Total input	578105	872447	796769	1092396	1412987	1788367
Total production	219510	344912	306223	417845	568116	682189
Estimated efficiency (% of production to input)	38	40	38	38	40	38

Statistics on electricity

British Virgin Islands

Item	2009	2010	2011	2012	2013	2014
Production, trade and consumption	**Million kilowatt-hours**					
Total main activity and autoproducer	*125	*128	131	*130	*130	*132
Combustible fuels	*125	*128	131	*130	*130	*132
Hydro	..	..	..	..	..	..
Nuclear	..	..	..	..	..	..
Other	..	..	..	..	..	..
Main activity	*125	*128	131	*130	*130	*132
Combustible fuels	*125	*128	131	*130	*130	*132
Hydro	..	..	..	..	..	..
Nuclear	..	..	..	..	..	..
Other	..	..	..	..	..	..
Own use in electricity, CHP and heat plants	*6	*6	6	*6	*6	*6
Net production	*119	*122	125	*124	*124	*126
Imports	..	..	..	..	..	..
Exports	..	..	..	..	..	..
Losses	*14	*15	*15	*15	*15	*15
Consumption	*105	*107	*110	*109	*109	*111
Energy industries own use	..	..	..	..	..	..
By industry and construction	*3	*3	*3	*3	*3	*4
By transport	..	..	..	..	..	..
By households and other cons.	*102	*104	*107	*106	*106	*107
Net installed capacity	**Thousand kilowatts**					
Total main activity and autoproducer	44	44	45	45	45	45
Combustible fuels	44	44	44	44	44	44
Hydro	..	..	..	..	..	..
Nuclear	..	..	..	..	..	..
Other	..	0	1	1	1	1
Main activity	44	44	45	45	45	45
Combustible fuels	44	44	44	44	44	44
Hydro	..	..	..	..	..	..
Nuclear	..	..	..	..	..	..
Other	..	0	1	1	1	1
Combustible fuel input	**Terajoules**					
Gas-diesel oil	*1350	*1380	*1415	*1406	*1406	*1410
Total input	*1350	*1380	*1415	*1406	*1406	*1410
Total production	*450	*461	472	*468	*468	*475
Estimated efficiency (% of production to input)	33	33	33	33	33	34

31

Statistics on electricity

Brunei Darussalam

Item	2009	2010	2011	2012	2013	2014
Production, trade and consumption	**Million kilowatt-hours**					
Total main activity and autoproducer	3612	3792	3725	3930	4402	4506
Combustible fuels	3612	3792	3723	3928	4400	4504
Hydro	..	..	..	..	..	..
Nuclear	..	..	..	..	..	..
Other	..	..	2	2	2	2
Main activity	3269	3440	3395	3524	3942	4010
Combustible fuels	3269	3440	3393	3522	3940	4008
Hydro	..	..	..	..	..	..
Nuclear	..	..	..	..	..	..
Other	..	..	2	2	2	2
Own use in electricity, CHP and heat plants	151	159	68	79	440	451
Net production	3461	3633	3657	3851	3962	4055
Imports	..	..	..	..	..	..
Exports	..	..	..	..	..	..
Losses	218	368	267	243	409	289
Consumption	3243	3265	3390	3609	3553	3765
Energy industries own use	372	351	326	401	398	408
By industry and construction	190	180	226	213	189	203
By transport	..	..	..	..	..	..
By households and other cons.	2681	2734	2838	2995	2966	3154
Net installed capacity	**Thousand kilowatts**					
Total main activity and autoproducer	*768	*785	*803	*826	*826	*844
Combustible fuels	*768	*785	*802	*825	*825	*843
Hydro	..	..	..	..	..	..
Nuclear	..	..	..	..	..	..
Other	..	..	*1	*1	*1	*1
Main activity	*700	*715	*731	*751	*751	*766
Combustible fuels	*700	*715	*730	*750	*750	*765
Hydro	..	..	..	..	..	..
Nuclear	..	..	..	..	..	..
Other	..	..	*1	*1	*1	*1
Combustible fuel input	**Terajoules**					
Gas-diesel oil	387	387	430	430	430	473
Natural gas	56153	54554	52528	55917	53434	55145
Total input	56540	54941	52958	56347	53864	55618
Total production	13003	13651	13403	14141	15840	16214
Estimated efficiency (% of production to input)	23	25	25	25	29	29

Statistics on electricity

Bulgaria

Item	2009	2010	2011	2012	2013	2014
Production, trade and consumption	Million kilowatt-hours					
Total main activity and autoproducer	42964	46653	50797	47329	43784	47485
Combustible fuels	23400	25001	29807	25518	22067	23858
Hydro	4053	5693	3691	3976	4795	5163
Nuclear	15256	15249	16314	15785	14171	15867
Other	255	710	985	2050	2751	2597
Main activity	42760	46411	50650	47059	43516	47232
Combustible fuels	23211	24773	29683	25263	21815	23619
Hydro	4053	5693	3691	3976	4795	5163
Nuclear	15256	15249	16314	15785	14171	15867
Other	240	696	962	2035	2735	2583
Own use in electricity, CHP and heat plants	4221	4435	4953	4470	3966	4254
Net production	38743	42218	45844	42859	39818	43231
Imports	2662	1167	1449	2353	3351	4319
Exports	7735	9613	12110	10661	9532	13774
Losses	4512	4480	4396	4231	3895	4013
Consumption	29094	29208	30762	30284	29741	29705
Energy industries own use	2247	2105	2341	2439	2209	2031
By industry and construction	8393	7818	8422	8267	8511	8706
By transport	467	399	367	302	277	306
By households and other cons.	17987	18886	19632	19276	18744	18662
Net installed capacity	Thousand kilowatts					
Total main activity and autoproducer	9597	10027	10236	11637	11604	11390
Combustible fuels	4369	4574	4541	4912	4701	4470
Hydro	3001	3048	3108	3129	3202	3219
Nuclear	1892	1892	1892	1906	1982	1975
Other	335	513	695	1690	1719	1726
Main activity	9553	9975	10190	11574	11539	11330
Combustible fuels	4325	4522	4495	4849	4636	4410
Hydro	3001	3048	3108	3129	3202	3219
Nuclear	1892	1892	1892	1906	1982	1975
Other	335	513	695	1690	1719	1726
Combustible fuel input	Terajoules					
Hard Coal	63426	64081	64770	44298	33215	39086
Brown Coal	171093	188950	246990	216856	181217	194391
Lignite briquettes	19551	17033	9132	10766	15517	16641
Biogas	8	67	83	4	81	412
Gas-diesel oil	43	1677	301	172	129	86
Fuel oil	5818	970	808	808	646	525
Refinery gas	990	644	248	743	743	347
Petroleum coke	1333	6110	3283	5395	6110	7150
Natural gas	31711	31882	35513	37004	36528	33056
Fuelwood	*26	*91	*316	*549	*757	*1279
Others	0	..	..	..	..	..
Total input	293999	311503	361443	316595	274942	292972
Total production	84240	90004	107305	91865	79441	85889
Estimated efficiency (% of production to input)	29	29	30	29	29	29

Statistics on electricity

Burkina Faso

Item	2009	2010	2011	2012	2013	2014
Production, trade and consumption	Million kilowatt-hours					
Total main activity and autoproducer	700	565	530	626	*731	*870
Combustible fuels	568	448	448	529	*625	*779
Hydro	132	118	82	97	106	90
Nuclear	..	..	..	..	..	..
Other	..	..	..	..	..	..
Main activity	700	565	442	479	*533	*621
Combustible fuels	568	448	360	382	*427	*530
Hydro	132	118	82	97	106	90
Nuclear	..	..	..	..	..	..
Other	..	..	..	..	..	..
Own use in electricity, CHP and heat plants	28	45	50	50	*50	*50
Net production	672	520	480	576	*681	*820
Imports	145	385	495	515	532	488
Exports	..	..	..	..	..	..
Losses	102	113	*115	*136	*165	*175
Consumption	714	*791	*860	*955	*1047	*1133
Energy industries own use	9	*10	*10	*10	*10	*10
By industry and construction	149	*165	*180	*200	*219	*260
By transport	10	*11	*12	*13	*15	*18
By households and other cons.	546	*605	*658	*732	*803	*845
Net installed capacity	Thousand kilowatts					
Total main activity and autoproducer	251	238	261	298	*298	*298
Combustible fuels	219	206	229	266	*266	*266
Hydro	32	32	32	32	*32	*32
Nuclear	..	..	..	..	..	..
Other	..	..	..	..	..	..
Main activity	251	238	200	237	*237	*237
Combustible fuels	219	206	168	205	*205	*205
Hydro	32	32	32	32	*32	*32
Nuclear	..	..	..	..	..	..
Other	..	..	..	..	..	..
Combustible fuel input	Terajoules					
Gas-diesel oil	*4945	*3978	*3939	*4506	*5214	*5246
Fuel oil	*566	*404	*404	*404	*404	*485
Total input	*5511	*4382	*4343	*4910	*5618	*5731
Total production	2043	1612	1613	1904	*2251	*2806
Estimated efficiency (% of production to input)	37	37	37	39	40	49

Statistics on electricity

Burundi

Item	2009	2010	2011	2012	2013	2014
Production, trade and consumption	**Million kilowatt-hours**					
Total main activity and autoproducer	121	142	141	142	159	174
Combustible fuels	4	18	13	3	21	34
Hydro	117	124	128	139	138	140
Nuclear	..	..	..	..	..	..
Other	..	..	..	..	..	..
Main activity	121	142	141	142	159	174
Combustible fuels	4	18	13	3	21	34
Hydro	117	124	128	139	138	140
Nuclear	..	..	..	..	..	..
Other	..	..	..	..	..	..
Own use in electricity, CHP and heat plants	8	8	8	8	8	8
Net production	113	134	133	134	151	166
Imports	*92	95	*99	*99	103	91
Exports	..	..	..	..	..	..
Losses	*34	*40	0	0	0	0
Consumption	*172	200	*244	*238	*259	*265
Energy industries own use	..	..	..	..	..	..
By industry and construction	*70	82	*100	*98	*106	*109
By transport	..	..	..	..	..	..
By households and other cons.	*102	118	*144	*140	*153	*156
Net installed capacity	**Thousand kilowatts**					
Total main activity and autoproducer	34	39	37	37	39	41
Combustible fuels	*2	*7	*5	*5	*7	*9
Hydro	32	32	32	32	32	32
Nuclear	..	..	..	..	..	..
Other	..	..	..	..	..	..
Main activity	34	39	37	37	39	41
Combustible fuels	*2	*7	*5	*5	*7	*9
Hydro	32	32	32	32	32	32
Nuclear	..	..	..	..	..	..
Other	..	..	..	..	..	..
Combustible fuel input	**Terajoules**					
Fuel oil	*61	*283	*202	*40	*242	*404
Total input	*61	*283	*202	*40	*242	*404
Total production	16	63	46	10	74	121
Estimated efficiency (% of production to input)	26	22	23	26	31	30

35

Cabo Verde

Item	2009	2010	2011	2012	2013	2014
Production, trade and consumption	**Million kilowatt-hours**					
Total main activity and autoproducer	309	347	365	375	404	409
Combustible fuels	305	343	341	306	321	338
Hydro	..	..	..	..	..	..
Nuclear	..	..	..	..	..	..
Other	5	4	25	69	83	71
Main activity	309	346	365	374	404	408
Combustible fuels	304	342	340	306	321	338
Hydro	..	..	..	..	..	..
Nuclear	..	..	..	..	..	..
Other	5	4	25	69	83	71
Own use in electricity, CHP and heat plants	11	11	12	11	10	11
Net production	298	336	353	364	394	398
Imports	..	..	..	..	..	..
Exports	..	..	..	..	..	..
Losses	77	84	89	97	102	103
Consumption	220	251	263	266	291	295
Energy industries own use	..	..	..	..	..	..
By industry and construction	22	20	18	23	37	33
By transport	..	..	..	..	..	..
By households and other cons.	198	231	245	243	255	262
Net installed capacity	**Thousand kilowatts**					
Total main activity and autoproducer	86	102	108	134	142	154
Combustible fuels	84	93	100	102	109	121
Hydro	..	..	..	..	..	..
Nuclear	..	..	..	..	..	..
Other	2	9	9	32	33	33
Main activity	85	101	108	134	142	154
Combustible fuels	83	93	99	101	108	121
Hydro	..	..	..	..	..	..
Nuclear	..	..	..	..	..	..
Other	2	9	9	32	33	33
Combustible fuel input	**Terajoules**					
Gas-diesel oil	722	925	856	778	602	615
Fuel oil	2214	2307	2367	2121	2275	2339
Total input	2936	3231	3223	2899	2877	2954
Total production	1097	1233	1227	1102	1156	1217
Estimated efficiency (% of production to input)	37	38	38	38	40	41

Statistics on electricity

Cambodia

Item	2009	2010	2011	2012	2013	2014
Production, trade and consumption	**Million kilowatt-hours**					
Total main activity and autoproducer	1256	994	1053	1434	1778	3059
Combustible fuels	1217	965	1005	914	759	1204
Hydro	37	26	45	517	1016	1852
Nuclear	..	..	..	..	..	..
Other	2	3	3	3	3	3
Main activity	1256	994	1053	1434	1778	3059
Combustible fuels	1217	965	1005	914	759	1204
Hydro	37	26	45	517	1016	1852
Nuclear	..	..	..	..	..	..
Other	2	3	3	3	3	3
Own use in electricity, CHP and heat plants	26	29	31	28	23	36
Net production	1230	965	1022	1406	1755	3023
Imports	725	1357	1644	1891	2050	1803
Exports	..	..	..	..	..	..
Losses	178	286	296	262	499	717
Consumption	1778	2038	2370	3035	3306	4109
Energy industries own use	..	..	..	..	..	..
By industry and construction	323	379	430	551	600	746
By transport	..	..	..	..	..	..
By households and other cons.	1455	1659	1940	2484	2706	3363
Net installed capacity	**Thousand kilowatts**					
Total main activity and autoproducer	373	362	571	584	1157	1513
Combustible fuels	359	347	362	357	472	582
Hydro	13	13	207	225	683	929
Nuclear	..	..	..	..	..	..
Other	*1	*2	2	*2	*2	*2
Main activity	373	362	571	584	1157	1513
Combustible fuels	359	347	362	357	472	582
Hydro	13	13	207	225	683	929
Nuclear	..	..	..	..	..	..
Other	*1	*2	2	*2	*2	*2
Combustible fuel input	**Terajoules**					
Brown Coal	318	338	378	418	1909	9745
Gas-diesel oil	2236	2580	2623	2365	1591	903
Fuel oil	10464	7272	7636	6868	4646	2626
Vegetal waste	268	288	288	288	158	201
Total input	13286	10478	10924	9939	8304	13475
Total production	4381	3474	3618	3290	2732	4334
Estimated efficiency (% of production to input)	33	33	33	33	33	32

Cameroon

Item	2009	2010	2011	2012	2013	2014
Production, trade and consumption	Million kilowatt-hours					
Total main activity and autoproducer	5783	5899	5943	6263	6849	6922
Combustible fuels	1767	1639	1546	1664	1994	1854
Hydro	4016	4260	4397	4599	4855	5068
Nuclear	..	..	..	..	..	..
Other	..	..	..	..	..	..
Main activity	4451	4630	4749	4968	5242	5473
Combustible fuels	435	370	352	369	387	405
Hydro	4016	4260	4397	4599	4855	5068
Nuclear	..	..	..	..	..	..
Other	..	..	..	..	..	..
Own use in electricity, CHP and heat plants	581	456	589	657	907	757
Net production	5202	5443	5354	5606	5942	6165
Imports	..	..	..	..	..	..
Exports	..	..	..	..	..	..
Losses	540	580	584	615	673	680
Consumption	4662	4863	4770	4991	5269	5485
Energy industries own use	..	..	..	..	..	..
By industry and construction	2720	2781	2573	2692	2842	3011
By transport	..	..	..	..	..	..
By households and other cons.	1942	2082	2197	2299	2427	2474
Net installed capacity	Thousand kilowatts					
Total main activity and autoproducer	1322	1307	1297	1348	1571	1549
Combustible fuels	603	*584	*568	*620	839	817
Hydro	719	723	728	728	732	*732
Nuclear	..	..	..	..	..	..
Other	..	..	..	..	..	..
Main activity	1022	1007	997	1048	1271	*1249
Combustible fuels	303	284	268	*320	539	517
Hydro	719	723	728	728	732	*732
Nuclear	..	..	..	..	..	..
Other	..	..	..	..	..	..
Combustible fuel input	Terajoules					
Gas-diesel oil	10148	11137	11352	11868	11911	6923
Fuel oil	2141	2303	2384	2505	2626	2788
Natural gas	4427	4466	3190	3828	6593	9571
Fuelwood	*669	*554	*577	*603	*637	*675
Total input	17385	18460	17503	18804	21767	19957
Total production	6361	5900	5566	5990	7178	6674
Estimated efficiency (% of production to input)	37	32	32	32	33	33

Statistics on electricity

Canada

Item	2009	2010	2011	2012	2013	2014
Production, trade and consumption	**Million kilowatt-hours**					
Total main activity and autoproducer	606346	595951	629901	632921	660795	656225
Combustible fuels	140817	144963	150042	146063	143406	139402
Hydro	368762	351461	375797	380340	391861	382574
Nuclear	90091	90658	93589	94862	103439	107678
Other	6676	8869	10473	11656	22089	26571
Main activity	557759	550062	577697	577950	604598	599532
Combustible fuels	123978	126283	128354	123130	118473	113890
Hydro	337016	324254	345282	348306	360615	351521
Nuclear	90091	90658	93589	94862	103439	107678
Other	6674	8867	10472	11652	22071	26443
Own use in electricity, CHP and heat plants	18031	18182	18938	18734	19484	19265
Net production	588315	577769	610963	614187	641311	636960
Imports	17493	18609	14398	10841	10694	12808
Exports	51152	43627	51076	57639	62578	58421
Losses	53096	51687	56000	53724	56326	58196
Consumption	503628	506162	516228	524189	527250	518977
Energy industries own use	27765	27524	28224	28598	29430	29738
By industry and construction	172675	175414	174897	171878	180525	180219
By transport	3855	3770	3852	4342	4705	4925
By households and other cons.	299333	299454	309255	319371	312590	304095
Net installed capacity	**Thousand kilowatts**					
Total main activity and autoproducer	131679	132375	132959	130554	133816	137344
Combustible fuels	40930	40424	38939	34660	35215	36217
Hydro	74687	75078	75573	75537	75537	75537
Nuclear	12665	12665	12665	13370	14033	14033
Other	3397	4208	5782	6987	9031	11557
Main activity	121262	121810	122465	120008	124562	127850
Combustible fuels	35576	34946	33532	29253	31337	32101
Hydro	69624	69991	70486	70398	70225	70232
Nuclear	12665	12665	12665	13370	14033	14033
Other	3397	4208	5782	6987	8967	11484
Combustible fuel input	**Terajoules**					
Hard Coal	148051	147275	109746	101195	96900	78913
Brown Coal	675038	649404	626937	562272	540679	539736
Biogas	8230	8200	10122	9097	9893	10206
Gas-diesel oil	9804	9847	9245	10234	9374	11223
Fuel oil	51308	30138	22180	20079	18059	22866
Refinery gas	13217	13464	13464	13514	18513	13662
Petroleum coke	27365	25123	22880	22523	28958	26325
Natural gas	444966	527270	601145	624612	615950	611028
Fuelwood	*37511	*48095	*47711	*39323	*39323	*40065
Municipal waste	2673	2673	2673	2910	2910	2910
Others	1325	1087	1248	1508	1190	1305
Total input	1419488	1462576	1467350	1407266	1381748	1358239
Total production	506941	521867	540151	525827	516262	501847
Estimated efficiency (% of production to input)	36	36	37	37	37	37

Statistics on electricity

Cayman Islands

Item	2009	2010	2011	2012	2013	2014
Production, trade and consumption	**Million kilowatt-hours**					
Total main activity and autoproducer	626	624	626	607	615	621
Combustible fuels	626	624	626	607	615	621
Hydro	..	..	..	..	..	..
Nuclear	..	..	..	..	..	..
Other	..	..	..	..	..	..
Main activity	626	624	626	607	615	621
Combustible fuels	626	624	626	607	615	621
Hydro	..	..	..	..	..	..
Nuclear	..	..	..	..	..	..
Other	..	..	..	..	..	..
Own use in electricity, CHP and heat plants	0	0	0	0	0	0
Net production	626	624	626	607	615	621
Imports	..	..	..	..	..	..
Exports	..	..	..	..	..	..
Losses	49	53	53	39	40	37
Consumption	576	571	573	568	575	584
Energy industries own use	..	..	..	..	..	..
By industry and construction	299	295	299	297	298	301
By transport	..	..	..	..	..	..
By households and other cons.	277	276	274	271	277	283
Net installed capacity	**Thousand kilowatts**					
Total main activity and autoproducer	153	151	151	150	150	132
Combustible fuels	153	151	151	150	150	132
Hydro	..	..	..	..	..	..
Nuclear	..	..	..	..	..	..
Other	..	..	..	..	..	..
Main activity	153	151	151	150	150	132
Combustible fuels	153	151	151	150	150	132
Hydro	..	..	..	..	..	..
Nuclear	..	..	..	..	..	..
Other	..	..	..	..	..	..
Combustible fuel input	**Terajoules**					
Gas-diesel oil	5289	5461	5547	5160	5074	5233
Total input	5289	5461	5547	5160	5074	5233
Total production	2254	2246	2254	2185	2214	2235
Estimated efficiency (% of production to input)	43	41	41	42	44	43

Statistics on electricity

Central African Rep.

Item	2009	2010	2011	2012	2013	2014
Production, trade and consumption	Million kilowatt-hours					
Total main activity and autoproducer	160	*160	*174	*177	*181	*183
Combustible fuels	*24	*25	*27	*27	*26	*26
Hydro	136	*135	*147	*150	*155	*157
Nuclear	..	..	..	..	..	..
Other	..	..	..	..	..	..
Main activity	160	*160	*174	*177	*181	*183
Combustible fuels	*24	*25	*27	*27	*26	*26
Hydro	136	*135	*147	*150	*155	*157
Nuclear	..	..	..	..	..	..
Other	..	..	..	..	..	..
Own use in electricity, CHP and heat plants	*2	*2	*9	*9	*11	*11
Net production	158	*158	*165	*168	*170	*172
Imports	..	..	..	..	..	..
Exports	..	..	..	..	..	..
Losses	*3	*9	*9	*9	*10	*10
Consumption	*148	*149	*156	*159	*161	*163
Energy industries own use	..	..	..	..	..	..
By industry and construction	*44	*44	*49	*50	*51	*52
By transport	..	..	..	..	..	..
By households and other cons.	*104	*105	*107	*109	*110	*111
Net installed capacity	Thousand kilowatts					
Total main activity and autoproducer	*44	*44	*44	*44	*44	*44
Combustible fuels	*19	*19	*19	*19	*19	*19
Hydro	*25	*25	*25	*25	*25	*25
Nuclear	..	..	..	..	..	..
Other	..	..	..	..	..	..
Main activity	*44	*44	*44	*44	*44	*44
Combustible fuels	*19	*19	*19	*19	*19	*19
Hydro	*25	*25	*25	*25	*25	*25
Nuclear	..	..	..	..	..	..
Other	..	..	..	..	..	..
Combustible fuel input	Terajoules					
Gas-diesel oil	*310	*323	*344	*344	*357	*361
Total input	*310	*323	*344	*344	*357	*361
Total production	*86	*90	*97	*97	*94	*94
Estimated efficiency (% of production to input)	28	28	28	28	26	26

Statistics on electricity

Chad

Item	2009	2010	2011	2012	2013	2014
Production, trade and consumption	**Million kilowatt-hours**					
Total main activity and autoproducer	183	200	*218	205	*225	*225
Combustible fuels	183	200	*218	205	*225	*225
Hydro	..	..	..	..	..	..
Nuclear	..	..	..	..	..	..
Other	..	..	..	..	..	..
Main activity	183	200	*218	205	*225	*225
Combustible fuels	183	200	*218	205	*225	*225
Hydro	..	..	..	..	..	..
Nuclear	..	..	..	..	..	..
Other	..	..	..	..	..	..
Own use in electricity, CHP and heat plants	16	18	*18	19	*19	*19
Net production	167	182	*200	186	*206	*206
Imports	..	..	..	..	..	..
Exports	..	..	..	..	..	..
Losses	..	..	..	..	..	..
Consumption	*167	*182	*200	*186	*206	*206
Energy industries own use	..	..	..	..	..	..
By industry and construction	*59	*64	*71	*66	*73	*73
By transport	..	..	..	..	..	..
By households and other cons.	*108	*118	*130	*120	*133	*133
Net installed capacity	**Thousand kilowatts**					
Total main activity and autoproducer	*41	*41	*41	*41	*47	*47
Combustible fuels	*41	*41	*41	*41	*47	*47
Hydro	..	..	..	..	..	..
Nuclear	..	..	..	..	..	..
Other	..	..	..	..	..	..
Main activity	*41	*41	*41	*41	*47	*47
Combustible fuels	*41	*41	*41	*41	*47	*47
Hydro	..	..	..	..	..	..
Nuclear	..	..	..	..	..	..
Other	..	..	..	..	..	..
Combustible fuel input	**Terajoules**					
Gas-diesel oil	*2025	*2215	*2408	*2408	*2460	*2460
Total input	*2025	*2215	*2408	*2408	*2460	*2460
Total production	659	720	*785	738	*810	*810
Estimated efficiency (% of production to input)	33	33	33	31	33	33

Statistics on electricity

Chile

Item	2009	2010	2011	2012	2013	2014
Production, trade and consumption	**Million kilowatt-hours**					
Total main activity and autoproducer	60722	60434	65713	69751	73065	73719
Combustible fuels	35250	38273	44366	49184	52602	48397
Hydro	25296	21717	21009	20158	19737	23099
Nuclear	..	..	..	..	..	..
Other	176	444	338	409	726	2223
Main activity	55186	56760	60286	63973	66643	67409
Combustible fuels	30308	35047	39001	43607	46374	42274
Hydro	24799	21381	20953	19979	19562	22940
Nuclear	..	..	..	..	..	..
Other	79	332	332	387	707	2195
Own use in electricity, CHP and heat plants	1993	2100	2524	3911	3321	1344
Net production	58729	58334	63189	65840	69744	72375
Imports	1348	958	732	0	0	0
Exports	..	..	..	..	..	..
Losses	6404	4967	4687	3499	4890	4820
Consumption	54533	55290	58492	62921	66040	67146
Energy industries own use	499	552	560	702	864	688
By industry and construction	36589	35853	38432	42154	42634	42260
By transport	422	432	478	474	511	1036
By households and other cons.	17023	18453	19022	19591	22031	23162
Net installed capacity	**Thousand kilowatts**					
Total main activity and autoproducer	15731	16231	17549	18153	18600	23400
Combustible fuels	10091	10576	11419	11957	12105	16049
Hydro	5452	5467	5946	5992	6094	6378
Nuclear	..	..	..	..	..	..
Other	188	188	184	204	401	973
Main activity	14420	14931	16358	17116	17305	22080
Combustible fuels	8893	9385	10244	10974	10988	14767
Hydro	5364	5383	5932	5950	6012	6350
Nuclear	..	..	..	..	..	..
Other	163	163	182	192	305	963
Combustible fuel input	**Terajoules**					
Hard Coal	126900	167144	197240	234552	285220	240980
Biogas	..	..	0	0	340	2241
Gas-diesel oil	80582	52374	40463	50654	26230	27348
Fuel oil	9373	8282	4969	8888	7474	6908
Liquefied petroleum gas	520	..	..	..	..	47
Refinery gas	1287	1287	396	495	2624	0
Petroleum coke	11440	11343	14365	16088	14203	6273
Natural gas	35190	81548	112011	112082	96694	81240
Fuelwood	30672	15373	25947	154961	176837	141213
Total input	295964	337350	395391	577720	609621	506250
Total production	126900	137783	159718	177062	189367	174229
Estimated efficiency (% of production to input)	43	41	40	31	31	34

Statistics on electricity

China

Item	2009	2010	2011	2012	2013	2014
Production, trade and consumption	**Million kilowatt-hours**					
Total main activity and autoproducer	3714650	4207160	4713019	4987553	5431637	5649583
Combustible fuels	2982780	3331928	3833702	3892814	4247009	4268649
Hydro	615640	722172	698945	872107	920291	1064337
Nuclear	70134	73880	86350	97394	111613	132538
Other	46096	79180	94022	125238	152724	184059
Main activity	3472747	3936941	4402106	4671846	5087205	5303396
Combustible fuels	2740877	3061709	3522789	3577107	3902577	3922462
Hydro	615640	722172	698945	872107	920291	1064337
Nuclear	70134	73880	86350	97394	111613	132538
Other	46096	79180	94022	125238	152724	184059
Own use in electricity, CHP and heat plants	282176	311927	381142	367045	404279	419079
Net production	3432474	3895233	4331877	4620508	5027358	5230504
Imports	6006	5545	6562	6874	7438	6750
Exports	17386	19059	19307	17653	18669	18158
Losses	225822	256824	270070	289616	314071	309988
Consumption	3195218	3624698	4048876	4319603	4701991	4909301
Energy industries own use	156977	174778	189065	197977	217886	222925
By industry and construction	2062664	2391972	2686060	2829423	3054959	3200445
By transport	61702	73453	84842	91537	100092	105924
By households and other cons.	913875	984495	1088909	1200666	1329054	1380007
Net installed capacity	**Thousand kilowatts**					
Total main activity and autoproducer	934600	1032150	1136211	1222980	1341077	1461939
Combustible fuels	705530	768829	834799	888257	945120	1005720
Hydro	202340	222633	240364	257090	288777	313981
Nuclear	9078	10820	12570	12570	14660	19880
Other	17652	29868	48478	65063	92520	122358
Main activity	874096	966418	1062368	1146783	1257710	1370728
Combustible fuels	651076	709670	768340	819680	870090	923630
Hydro	196290	216060	232980	249470	280440	304860
Nuclear	9078	10820	12570	12570	14660	19880
Other	17652	29868	48478	65063	92520	122358
Combustible fuel input	**Terajoules**					
Hard Coal	29772429	31363474	36239409	36027135	38352359	36886601
Fuel oil	37293	131381	98818	93094	86630	14362
Refinery gas	120038	123651	103336	102366	93035	112815
Petroleum coke	32874	35578	39540	42455	34983	37161
Natural gas	595936	797850	947506	981657	1061128	1137996
Gasworks gas	246026	36686	72887	76274	82138	85229
Coke-oven gas	159093	189789	213395	238945	274931	299059
Blast furnace gas	..	298322	350393	362066	468351	539699
Municipal waste	110854	111031	127686	153223	189996	229758
Industrial waste	..	130509	155076	157922	177157	186570
Others	26751	71762	54795	30823	33974	31211
Total input	31101294	33290032	38402842	38265960	40854682	39560462
Total production	10738008	11994942	13801327	14014130	15289234	15367138
Estimated efficiency (% of production to input)	35	36	36	37	37	39

Statistics on electricity

China, Hong Kong SAR

Item	2009	2010	2011	2012	2013	2014
Production, trade and consumption	**Million kilowatt-hours**					
Total main activity and autoproducer	38728	38292	39026	38752	39063	39803
Combustible fuels	38728	38292	39026	38752	39063	39803
Hydro	..	..	..	..	..	..
Nuclear	..	..	..	..	..	..
Other	..	..	..	..	..	..
Main activity	38728	38292	39026	38752	39063	39803
Combustible fuels	38728	38292	39026	38752	39063	39803
Hydro	..	..	..	..	..	..
Nuclear	..	..	..	..	..	..
Other	..	..	..	..	..	..
Own use in electricity, CHP and heat plants	0	0	0	0	0	0
Net production	38728	38292	39026	38752	39063	39803
Imports	10963	10511	10735	11156	9969	10288
Exports	3731	2609	2957	1838	1650	1226
Losses	4469	4331	4739	5039	4826	4986
Consumption	41491	41863	42065	43031	42556	43880
Energy industries own use	..	..	..	..	..	..
By industry and construction	3095	3078	3084	3134	3108	3134
By transport	..	..	..	..	..	..
By households and other cons.	38396	38785	38981	39897	39448	40746
Net installed capacity	**Thousand kilowatts**					
Total main activity and autoproducer	12624	12624	12624	12625	12625	12625
Combustible fuels	12624	12624	12624	12625	12625	12625
Hydro	..	..	..	..	..	..
Nuclear	..	..	..	..	..	..
Other	..	..	..	..	..	..
Main activity	12624	12624	12624	12625	12625	12625
Combustible fuels	12624	12624	12624	12625	12625	12625
Hydro	..	..	..	..	..	..
Nuclear	..	..	..	..	..	..
Other	..	..	..	..	..	..
Combustible fuel input	**Terajoules**					
Hard Coal	299510	260615	309976	305150	325274	340806
Gas-diesel oil	731	602	559	4085	602	2064
Fuel oil	1131	929	970	3798	1212	40
Natural gas	87636	114576	89083	82075	79589	73829
Total input	389009	376722	400588	395108	406677	416739
Total production	139421	137851	140494	139507	140627	143291
Estimated efficiency (% of production to input)	36	37	35	35	35	34

Statistics on electricity

China, Macao SAR

Item	2009	2010	2011	2012	2013	2014
Production, trade and consumption	**Million kilowatt-hours**					
Total main activity and autoproducer	1466	1077	886	561	413	641
Combustible fuels	1466	1077	886	561	413	641
Hydro	..	..	..	..	..	..
Nuclear	..	..	..	..	..	..
Other	..	..	..	..	..	..
Main activity	1350	944	733	392	227	435
Combustible fuels	1350	944	733	392	227	435
Hydro	..	..	..	..	..	..
Nuclear	..	..	..	..	..	..
Other	..	..	..	..	..	..
Own use in electricity, CHP and heat plants	61	48	52	44	35	42
Net production	1405	1029	834	517	378	599
Imports	2227	2786	3165	3855	4059	4099
Exports	..	..	..	..	..	..
Losses	169	160	142	167	146	166
Consumption	3463	3656	3858	4205	4290	4532
Energy industries own use	3	3	3	2	2	2
By industry and construction	337	214	203	213	221	251
By transport	..	..	..	..	..	..
By households and other cons.	3123	3439	3652	3990	4067	4279
Net installed capacity	**Thousand kilowatts**					
Total main activity and autoproducer	497	497	497	497	497	497
Combustible fuels	497	497	497	497	497	497
Hydro	..	..	..	..	..	..
Nuclear	..	..	..	..	..	..
Other	..	..	..	..	..	..
Main activity	472	472	472	472	472	472
Combustible fuels	472	472	472	472	472	472
Hydro	..	..	..	..	..	..
Nuclear	..	..	..	..	..	..
Other	..	..	..	..	..	..
Combustible fuel input	**Terajoules**					
Gas-diesel oil	43	0	43	0	0	0
Fuel oil	7474	2464	3394	2909	1778	1737
Natural gas	3635	6036	2872	0	0	2259
Municipal waste	*1670	*1915	*1541	*1627	*1842	*2058
Total input	12822	10415	7850	4536	*3620	6054
Total production	5278	3877	3190	2020	1487	2308
Estimated efficiency (% of production to input)	41	37	41	45	41	38

Statistics on electricity

Colombia

Item	2009	2010	2011	2012	2013	2014
Production, trade and consumption	**Million kilowatt-hours**					
Total main activity and autoproducer	62574	60234	66472	68344	70858	70059
Combustible fuels	21530	19669	17514	20689	26423	25149
Hydro	40986	40527	48917	47600	44377	44852
Nuclear	..	..	..	..	..	..
Other	58	38	41	55	58	58
Main activity	59357	56794	62881	64554	67394	66316
Combustible fuels	18526	16452	14079	17031	23080	21545
Hydro	40773	40304	48761	47468	44256	44713
Nuclear	..	..	..	..	..	..
Other	58	38	41	55	58	58
Own use in electricity, CHP and heat plants	1720	1749	1802	1844	1912	1977
Net production	60854	58485	64670	66500	68946	68082
Imports	21	10	8	7	5	47
Exports	1074	796	1540	713	1374	849
Losses	8387	9115	7412	7700	6660	7487
Consumption	53425	54701	57845	59285	67260	51268
Energy industries own use	..	..	..	..	..	..
By industry and construction	20206	20436	22833	22626	25434	16374
By transport	60	62	63	65	75	77
By households and other cons.	33159	34203	34949	36594	41751	34817
Net installed capacity	**Thousand kilowatts**					
Total main activity and autoproducer	14009	14747	15041	15033	15177	14823
Combustible fuels	4918	5376	5158	5088	5131	5043
Hydro	9040	9300	9810	9870	9967	9696
Nuclear	..	..	..	..	..	..
Other	51	71	73	75	79	84
Main activity	13508	14246	14427	14413	14557	14203
Combustible fuels	4460	4918	4636	4560	4603	4515
Hydro	8997	9257	9718	9778	9875	9604
Nuclear	..	..	..	..	..	..
Other	51	71	73	75	79	84
Combustible fuel input	**Terajoules**					
Hard Coal	31955	32874	34091	32841	34290	71029
Gas-diesel oil	2623	3999	43	1763	1849	1978
Fuel oil	1212	2182	1616	2384	3515	1374
Natural gas	121938	133080	103376	108401	132546	156632
Coke-oven gas	842	664	590	590	486	..
Blast furnace gas	842	*664	501	590	525	400
Bagasse	15406	25251	27854	30672	31568	13805
Vegetal waste	958	716	609	661	572	..
Total input	175775	199430	168680	177902	205350	245217
Total production	77508	70808	63050	74480	95123	90536
Estimated efficiency (% of production to input)	44	36	37	42	46	37

Statistics on electricity

Comoros

Item	2009	2010	2011	2012	2013	2014
Production, trade and consumption	**Million kilowatt-hours**					
Total main activity and autoproducer	*46	*64	*57	*58	*56	*55
Combustible fuels	*46	*64	*57	*58	*56	*55
Hydro	..	..	..	..	..	..
Nuclear	..	..	..	..	..	..
Other	..	..	..	..	..	..
Main activity	*46	*64	*57	*58	*56	*55
Combustible fuels	*46	*64	*57	*58	*56	*55
Hydro	..	..	..	..	..	..
Nuclear	..	..	..	..	..	..
Other	..	..	..	..	..	..
Own use in electricity, CHP and heat plants	0	0	0	0	0	0
Net production	*46	*64	*57	*58	*56	*55
Imports	..	..	..	..	..	..
Exports	..	..	..	..	..	..
Losses	*10	*13	*12	*12	*12	*11
Consumption	*37	*50	*46	*46	*44	*44
Energy industries own use	..	..	..	..	..	..
By industry and construction	..	..	..	..	..	..
By transport	..	..	..	..	..	..
By households and other cons.	*37	*50	*46	*46	*44	*44
Net installed capacity	**Thousand kilowatts**					
Total main activity and autoproducer	*23	*23	*23	*23	*24	*25
Combustible fuels	*23	*23	*23	*23	*24	*25
Hydro	..	..	..	..	..	..
Nuclear	..	..	..	..	..	..
Other	..	..	..	..	..	..
Main activity	*23	*23	*23	*23	*24	*25
Combustible fuels	*23	*23	*23	*23	*24	*25
Hydro	..	..	..	..	..	..
Nuclear	..	..	..	..	..	..
Other	..	..	..	..	..	..
Combustible fuel input	**Terajoules**					
Gas-diesel oil	*503	*688	*619	*632	*606	*589
Total input	*503	*688	*619	*632	*606	*589
Total production	*167	*229	*207	*210	*202	*197
Estimated efficiency (% of production to input)	33	33	33	33	33	33

Statistics on electricity

Congo

Item	2009	2010	2011	2012	2013	2014
Production, trade and consumption	**Million kilowatt-hours**					
Total main activity and autoproducer	539	784	1293	1700	1709	1740
Combustible fuels	209	355	502	713	738	788
Hydro	330	429	791	987	971	952
Nuclear	..	..	..	..	..	..
Other	..	..	..	..	..	..
Main activity	353	441	800	987	971	952
Combustible fuels	23	12	9	0	0	0
Hydro	330	429	791	987	971	952
Nuclear	..	..	..	..	..	..
Other	..	..	..	..	..	..
Own use in electricity, CHP and heat plants	95	88	145	191	192	195
Net production	444	696	1148	1509	1517	1545
Imports	440	281	27	25	56	18
Exports	..	0	0	0	19	22
Losses	379	466	671	757	761	775
Consumption	481	500	606	792	802	798
Energy industries own use	..	..	..	..	..	..
By industry and construction	237	246	283	370	375	373
By transport	..	..	..	..	..	..
By households and other cons.	244	254	323	422	427	425
Net installed capacity	**Thousand kilowatts**					
Total main activity and autoproducer	*238	*238	*369	*390	*390	*390
Combustible fuels	*119	*119	*149	*170	*170	*170
Hydro	*119	*119	*220	*220	*220	*220
Nuclear	..	..	..	..	..	..
Other	..	..	..	..	..	..
Main activity	*148	*148	*249	*230	*230	*230
Combustible fuels	*29	*29	*29	*10	*10	*10
Hydro	*119	*119	*220	*220	*220	*220
Nuclear	..	..	..	..	..	..
Other	..	..	..	..	..	..
Combustible fuel input	**Terajoules**					
Gas-diesel oil	215	129	129	0	0	0
Fuel oil	121	40	0	0	0	0
Natural gas	2125	3923	5734	8293	8584	9166
Total input	2461	4092	5863	8293	8584	9166
Total production	752	1278	1807	2567	2657	2837
Estimated efficiency (% of production to input)	31	31	31	31	31	31

Statistics on electricity

Cook Islands

Item	2009	2010	2011	2012	2013	2014
Production, trade and consumption	**Million kilowatt-hours**					
Total main activity and autoproducer	33	34	34	33	33	32
Combustible fuels	33	34	34	33	33	32
Hydro	..	..	..	..	..	..
Nuclear	..	..	..	..	..	..
Other	..	..	..	..	..	..
Main activity	33	34	34	33	33	32
Combustible fuels	33	34	34	33	33	32
Hydro	..	..	..	..	..	..
Nuclear	..	..	..	..	..	..
Other	..	..	..	..	..	..
Own use in electricity, CHP and heat plants	0	0	0	0	0	0
Net production	33	34	34	33	33	32
Imports	..	..	..	..	..	..
Exports	..	..	..	..	..	..
Losses	..	..	..	..	..	..
Consumption	33	34	34	33	33	32
Energy industries own use	..	..	..	..	..	..
By industry and construction	..	..	..	..	..	..
By transport	..	..	..	..	..	..
By households and other cons.	33	34	34	33	33	32
Net installed capacity	**Thousand kilowatts**					
Total main activity and autoproducer	*8	*8	*8	*8	*8	*8
Combustible fuels	*8	*8	*8	*8	*8	*8
Hydro	..	..	..	..	..	..
Nuclear	..	..	..	..	..	..
Other	..	..	..	..	..	..
Main activity	*8	*8	*8	*8	*8	*8
Combustible fuels	*8	*8	*8	*8	*8	*8
Hydro	..	..	..	..	..	..
Nuclear	..	..	..	..	..	..
Other	..	..	..	..	..	..
Combustible fuel input	**Terajoules**					
Gas-diesel oil	*387	*387	*387	*387	*387	*387
Total input	*387	*387	*387	*387	*387	*387
Total production	118	121	121	118	117	114
Estimated efficiency (% of production to input)	30	31	31	31	30	30

Statistics on electricity

Costa Rica

Item	2009	2010	2011	2012	2013	2014
Production, trade and consumption	Million kilowatt-hours					
Total main activity and autoproducer	9311	9583	9831	10174	10235	10217
Combustible fuels	575	786	1002	1009	1379	1224
Hydro	7224	7262	7135	7233	6851	6717
Nuclear	..	..	..	..	..	..
Other	1512	1535	1694	1932	2005	2276
Main activity	7441	7542	7837	7988	10052	10036
Combustible fuels	451	641	863	830	1196	1043
Hydro	5936	5872	5831	5877	6851	6717
Nuclear	..	..	..	..	..	..
Other	1054	1028	1143	1281	2005	2276
Own use in electricity, CHP and heat plants	68	101	91	109	109	117
Net production	9244	9483	9741	10066	10126	10100
Imports	151	165	283	419	531	759
Exports	134	136	320	402	484	555
Losses	986	970	1060	1083	1083	1106
Consumption	8313	8565	8673	9019	9103	9207
Energy industries own use	..	..	..	..	..	..
By industry and construction	1753	1828	1818	1893	1898	1858
By transport	..	..	..	..	..	..
By households and other cons.	6560	6736	6855	7126	7205	7349
Net installed capacity	Thousand kilowatts					
Total main activity and autoproducer	2413	2794	2889	2724	2810	3215
Combustible fuels	615	911	862	656	634	618
Hydro	1532	1596	1682	1700	1729	2045
Nuclear	..	..	..	..	..	..
Other	266	287	345	368	*448	552
Main activity	2102	2438	2521	2354	2440	2845
Combustible fuels	595	882	834	635	613	597
Hydro	1337	1386	1472	1490	1519	1835
Nuclear	..	..	..	..	..	..
Other	170	170	215	229	309	413
Combustible fuel input	Terajoules					
Biogas	..	..	..	..	1	1
Gas-diesel oil	4327	6100	3575	850	2666	2580
Fuel oil	702	1113	4740	5655	8120	7151
Bagasse	802	936	883	1196	1196	1154
Vegetal waste	108	107	109	118	118	114
Total input	5939	8256	9307	7819	12101	11000
Total production	2068	2830	3608	3632	4964	4406
Estimated efficiency (% of production to input)	35	34	39	46	41	40

Côte d'Ivoire

Item	2009	2010	2011	2012	2013	2014
Production, trade and consumption	**Million kilowatt-hours**					
Total main activity and autoproducer	5871	5965	6099	7016	7650	8286
Combustible fuels	3740	4347	4325	5227	6044	6373
Hydro	2131	1618	1774	1789	1606	1913
Nuclear	..	..	..	..	..	..
Other	..	..	..	..	..	..
Main activity	5797	5884	6034	6948	7580	8214
Combustible fuels	3666	4266	4260	5159	5974	6301
Hydro	2131	1618	1774	1789	1606	1913
Nuclear	..	..	..	..	..	..
Other	..	..	..	..	..	..
Own use in electricity, CHP and heat plants	294	298	305	351	383	414
Net production	5577	5667	5794	6665	7267	7872
Imports	..	145	22	54	0	0
Exports	484	491	615	645	820	877
Losses	1365	1204	1365	1366	1680	1187
Consumption	3728	4117	3836	4708	4767	5808
Energy industries own use	..	..	..	..	..	..
By industry and construction	967	1068	1312	1640	1755	1738
By transport	..	..	..	..	..	..
By households and other cons.	2761	3049	2524	3068	3012	4070
Net installed capacity	**Thousand kilowatts**					
Total main activity and autoproducer	*1414	*1414	*1414	*1414	*1525	*1655
Combustible fuels	*823	*843	*843	*1043	*1154	*1223
Hydro	*591	*571	*571	*371	*371	*432
Nuclear	..	..	..	..	..	..
Other	..	..	..	..	..	..
Main activity	*1391	*1391	*1391	*1391	*1502	*1632
Combustible fuels	*800	*820	*820	*1020	*1131	*1200
Hydro	*591	*571	*571	*371	*371	*432
Nuclear	..	..	..	..	..	..
Other	..	..	..	..	..	..
Combustible fuel input	**Terajoules**					
Gas-diesel oil	86	86	86	129	129	129
Fuel oil	40	970	323	5090	2303	6141
Natural gas	45515	53123	52420	60590	64384	64991
Vegetal waste	1537	1471	1364	1401	1439	1478
Total input	47178	55650	54193	67210	68255	72739
Total production	13464	15649	15570	18817	21758	22943
Estimated efficiency (% of production to input)	29	28	29	28	32	32

Statistics on electricity

Croatia

Item	2009	2010	2011	2012	2013	2014
Production, trade and consumption	**Million kilowatt-hours**					
Total main activity and autoproducer	13455	14902	11373	10755	14052	13554
Combustible fuels	5907	5531	6010	5425	4797	3664
Hydro	7494	9232	5162	4999	8727	9125
Nuclear	..	..	..	..	..	..
Other	54	139	201	331	528	765
Main activity	13050	14447	10857	10367	13720	13209
Combustible fuels	5507	5084	5498	5042	4470	3326
Hydro	7489	9224	5158	4994	8722	9118
Nuclear	..	..	..	..	..	..
Other	54	139	201	331	528	765
Own use in electricity, CHP and heat plants	422	470	422	359	392	395
Net production	13033	14432	10951	10396	13660	13159
Imports	11892	12415	13985	13174	11260	10898
Exports	6889	8447	6830	5743	7391	6945
Losses	2019	2022	1831	1887	1944	1764
Consumption	16017	16378	16275	15940	15585	15348
Energy industries own use	506	516	540	590	513	515
By industry and construction	3392	3478	3349	3039	3148	3291
By transport	267	267	261	249	238	231
By households and other cons.	11852	12117	12125	12062	11686	11311
Net installed capacity	**Thousand kilowatts**					
Total main activity and autoproducer	3976	4010	4015	4222	4312	4428
Combustible fuels	1876	1781	1787	1897	1849	1933
Hydro	2083	2113	2110	2141	2190	2114
Nuclear	..	..	..	..	..	..
Other	17	116	118	184	273	381
Main activity	3762	3796	3801	4021	4155	4271
Combustible fuels	1666	1571	1577	1698	1694	1778
Hydro	2079	2109	2106	2139	2188	2112
Nuclear	..	..	..	..	..	..
Other	17	116	118	184	273	381
Combustible fuel input	**Terajoules**					
Hard Coal	15744	22204	23347	21147	23519	22796
Brown Coal	18	175	176	120	123	143
Biogas	195	260	276	437	655	974
Gas-diesel oil	86	43	43	86	43	43
Fuel oil	20240	5898	7393	5414	2222	1535
Liquefied petroleum gas	0	0	0	47	0	..
Refinery gas	0	50	50	149	99	99
Petroleum coke	0	0	33	0	0	..
Natural gas	25603	28028	28364	28239	24563	15770
Fuelwood	29	24	833	1046	1146	1190
Total input	61915	56682	60514	56684	52370	42550
Total production	21265	19912	21636	19530	17269	13190
Estimated efficiency (% of production to input)	34	35	36	34	33	31

Statistics on electricity

Cuba

Item	2009	2010	2011	2012	2013	2014
Production, trade and consumption	**Million kilowatt-hours**					
Total main activity and autoproducer	17727	17387	17759	18432	19140	19366
Combustible fuels	17576	17290	17660	18321	19012	19262
Hydro	151	97	99	111	127	104
Nuclear	..	..	..	..	..	..
Other	..	..	..	..	..	..
Main activity	16859	16584	16944	17596	18263	18529
Combustible fuels	16708	16487	16845	17485	18135	18424
Hydro	151	97	99	111	127	104
Nuclear	..	..	..	..	..	..
Other	..	..	..	..	..	..
Own use in electricity, CHP and heat plants	886	903	947	982	1018	1056
Net production	16841	16484	16812	17450	18121	18310
Imports	..	..	..	..	..	..
Exports	..	..	..	..	..	..
Losses	2749	2788	2787	2919	2943	2961
Consumption	16817	17109	17760	18432	18135	16405
Energy industries own use	..	..	..	..	..	..
By industry and construction	3703	3782	4781	4900	4898	4622
By transport	264	251	254	258	296	302
By households and other cons.	12850	13076	12725	13274	12941	11480
Net installed capacity	**Thousand kilowatts**					
Total main activity and autoproducer	5551	5853	5914	5699	6055	6169
Combustible fuels	5493	5790	5849	5637	5992	6106
Hydro	58	63	65	62	63	63
Nuclear	..	..	..	..	..	..
Other	..	..	..	..	..	..
Main activity	5004	5304	5370	5180	5585	5671
Combustible fuels	4946	5241	5305	5118	5522	5608
Hydro	58	63	65	62	63	63
Nuclear	..	..	..	..	..	..
Other	..	..	..	..	..	..
Combustible fuel input	**Terajoules**					
Crude oil	*111799	*110022	*108034	*102197	*98352	*96503
Gas-diesel oil	*25241	*19823	*19307	*18748	*19780	*18877
Fuel oil	*24200	*73932	*73932	*84961	*64680	*61933
Natural gas	*15774	*14654	*13930	*14135	*14559	*14996
Bagasse	*28286	*23274	*29691	*28896	*28000	*34315
Total input	*205300	*241706	*244895	*248937	*225371	*226625
Total production	63274	62244	63576	65956	68444	69343
Estimated efficiency (% of production to input)	31	26	26	26	30	31

Statistics on electricity

Curaçao

Item	2009	2010	2011	2012	2013	2014
Production, trade and consumption			Million kilowatt-hours			
Total main activity and autoproducer	..	..	..	917	904	891
Combustible fuels	..	..	..	885	872	859
Hydro	..	..	..	..	..	..
Nuclear	..	..	..	..	..	..
Other	..	..	..	32	32	32
Main activity	..	..	..	431	425	419
Combustible fuels	..	..	..	399	393	387
Hydro	..	..	..	..	..	..
Nuclear	..	..	..	..	..	..
Other	..	..	..	32	32	32
Own use in electricity, CHP and heat plants	..	..	..	86	85	84
Net production	..	..	..	831	819	807
Imports	..	..	..	..	..	..
Exports	..	..	..	..	..	..
Losses	..	..	..	147	145	143
Consumption	..	..	..	684	674	664
Energy industries own use	..	..	..	..	..	..
By industry and construction	..	..	..	376	371	365
By transport	..	..	..	..	..	..
By households and other cons.	..	..	..	308	303	299
Net installed capacity			Thousand kilowatts			
Total main activity and autoproducer	..	..	..	*265	*265	*265
Combustible fuels	..	..	..	*235	*235	*235
Hydro	..	..	..	..	..	..
Nuclear	..	..	..	..	..	..
Other	..	..	..	*30	*30	*30
Main activity	..	..	..	*132	*132	*132
Combustible fuels	..	..	..	*102	*102	*102
Hydro	..	..	..	..	..	..
Nuclear	..	..	..	..	..	..
Other	..	..	..	*30	*30	*30
Combustible fuel input			Terajoules			
Fuel oil	..	..	..	8201	8080	7959
Total input	..	..	..	8201	8080	7959
Total production	..	..	..	3186	3139	3092
Estimated efficiency (% of production to input)	..	..	..	39	39	39

Statistics on electricity

Cyprus

Item	2009	2010	2011	2012	2013	2014
Production, trade and consumption	**Million kilowatt-hours**					
Total main activity and autoproducer	5215	5322	4929	4717	4290	4350
Combustible fuels	5211	5284	4803	4510	4012	4084
Hydro	..	..	..	..	..	..
Nuclear	..	..	..	..	..	..
Other	4	38	126	207	278	266
Main activity	5138	5246	4861	4657	4230	4272
Combustible fuels	5135	5210	4737	4452	3954	4026
Hydro	..	..	..	..	..	..
Nuclear	..	..	..	..	..	..
Other	3	36	124	205	276	246
Own use in electricity, CHP and heat plants	266	222	231	174	171	205
Net production	4949	5100	4698	4543	4119	4145
Imports	..	..	..	..	..	..
Exports	..	..	..	..	..	..
Losses	189	220	160	137	186	173
Consumption	4758	4889	4729	4417	3935	3972
Energy industries own use	7	7	8	8	14	7
By industry and construction	591	579	529	485	462	463
By transport	..	..	..	..	..	..
By households and other cons.	4160	4303	4192	3924	3459	3502
Net installed capacity	**Thousand kilowatts**					
Total main activity and autoproducer	1426	1561	1736	1726	1700	1734
Combustible fuels	1421	1471	1591	1561	1517	1513
Hydro	..	..	..	..	..	..
Nuclear	..	..	..	..	..	..
Other	5	90	145	165	183	221
Main activity	1393	1528	1699	1689	1663	1692
Combustible fuels	1389	1439	1555	1525	1481	1481
Hydro	..	..	..	..	..	..
Nuclear	..	..	..	..	..	..
Other	4	89	144	164	182	211
Combustible fuel input	**Terajoules**					
Biogas	144	169	272	283	289	292
Gas-diesel oil	3956	6794	4902	9288	10234	5375
Fuel oil	47510	42945	42905	36279	26381	32199
Total input	51610	49908	48079	45850	36904	37866
Total production	18760	19022	17291	16236	14443	14702
Estimated efficiency (% of production to input)	36	38	36	35	39	39

Czechia

Item	2009	2010	2011	2012	2013	2014
Production, trade and consumption	Million kilowatt-hours					
Total main activity and autoproducer	82250	85910	87561	87573	87065	86024
Combustible fuels	51683	53581	54035	51824	50167	50138
Hydro	2982	3380	2664	2860	3639	2961
Nuclear	27208	27998	28283	30324	30745	30325
Other	377	951	2579	2565	2514	2600
Main activity	73344	76596	79432	79588	78847	77546
Combustible fuels	43347	44953	46409	44362	42570	42224
Hydro	2412	2694	2161	2337	3018	2397
Nuclear	27208	27998	28283	30324	30745	30325
Other	377	951	2579	2565	2514	2600
Own use in electricity, CHP and heat plants	6260	6446	6533	6485	6207	6118
Net production	75990	79464	81028	81088	80858	79906
Imports	8586	6642	10457	11587	10571	11842
Exports	22230	21590	27501	28707	27458	28142
Losses	4487	4466	4405	4187	4098	3847
Consumption	57852	60042	59570	59773	59865	59743
Energy industries own use	2946	2838	2819	3119	3174	3540
By industry and construction	21821	22592	23216	22698	23204	23005
By transport	2045	1548	1633	1634	1615	1581
By households and other cons.	31040	33064	31902	32322	31872	31617
Net installed capacity	Thousand kilowatts					
Total main activity and autoproducer	18326	19829	20181	20447	21079	21970
Combustible fuels	11654	11793	11888	11915	12211	13082
Hydro	2184	2196	2197	2212	2252	2252
Nuclear	3830	3900	3970	4040	4290	4290
Other	658	1940	2126	2280	2326	2346
Main activity	16137	17968	18294	18554	19178	20090
Combustible fuels	9626	10105	10175	10205	10498	11392
Hydro	2023	2023	2023	2029	2064	2062
Nuclear	3830	3900	3970	4040	4290	4290
Other	658	1940	2126	2280	2326	2346
Combustible fuel input	Terajoules					
Hard Coal	94317	103651	98082	86221	92063	85581
Brown Coal	472902	486235	479439	458264	426270	411735
Biogas	3874	5114	7514	11552	18284	19693
Natural gas	18651	18829	21673	21705	21510	23855
Gasworks gas	16642	18152	17741	16801	16269	17157
Coke-oven gas	4865	5190	6711	6278	6371	6391
Blast furnace gas	6034	8883	7891	7130	8585	8761
Fuelwood	*13382	*15457	*17370	*19150	*21434	*24091
Municipal waste	1390	2906	4141	4410	4201	4251
Other recovered gases	267	621	654	656	637	1143
Others	4610	3561	2667	2215	1108	855
Total input	636934	668598	663883	634382	616732	603513
Total production	186059	192892	194526	186566	180601	180497
Estimated efficiency (% of production to input)	29	29	29	29	29	30

Dem. Rep. of the Congo

Item	2009	2010	2011	2012	2013	2014
Production, trade and consumption	**Million kilowatt-hours**					
Total main activity and autoproducer	7854	7905	7899	7595	8240	8831
Combustible fuels	86	86	88	9	9	11
Hydro	7768	7819	7811	7586	8231	8820
Nuclear	..	..	..	..	..	..
Other	..	..	..	..	..	..
Main activity	7526	7559	7535	7595	8240	8824
Combustible fuels	6	6	8	9	9	4
Hydro	7520	7553	7527	7586	8231	8820
Nuclear	..	..	..	..	..	..
Other	..	..	..	..	..	..
Own use in electricity, CHP and heat plants	15	494	493	480	521	640
Net production	7839	7411	7406	7115	7719	8191
Imports	105	161	44	626	626	1138
Exports	887	916	171	48	48	69
Losses	403	393	561	430	467	1895
Consumption	6654	6263	6718	7383	7252	7899
Energy industries own use	..	..	..	..	..	..
By industry and construction	4218	3970	4258	3652	3243	4342
By transport	..	..	..	15	0	0
By households and other cons.	2436	2293	2460	3716	4009	3557
Net installed capacity	**Thousand kilowatts**					
Total main activity and autoproducer	*2476	*2506	*2506	*2506	*2606	*2606
Combustible fuels	*34	*34	*34	*34	*34	*34
Hydro	*2442	*2472	*2472	*2472	*2572	*2572
Nuclear	..	..	..	..	..	..
Other	..	..	..	..	..	..
Main activity	*1899	*1929	*1929	*1929	*2029	*2029
Combustible fuels	*30	*30	*30	*30	*30	*30
Hydro	*1869	*1899	*1899	*1899	*1999	*1999
Nuclear	..	..	..	..	..	..
Other	..	..	..	..	..	..
Combustible fuel input	**Terajoules**					
Gas-diesel oil	86	86	86	43	43	43
Natural gas	914	914	914	0	0	80
Fuelwood	..	..	..	5032	5032	2784
Total input	1000	1000	1000	5075	5075	2907
Total production	310	310	317	32	32	40
Estimated efficiency (% of production to input)	31	31	32	1	1	1

Statistics on electricity

Denmark

Item	2009	2010	2011	2012	2013	2014
Production, trade and consumption	**Million kilowatt-hours**					
Total main activity and autoproducer	36383	38862	35229	30701	34760	32183
Combustible fuels	29639	31026	25423	20310	23106	18493
Hydro	19	21	17	17	13	15
Nuclear	..	..	..	..	..	..
Other	6725	7815	9789	10374	11641	13675
Main activity	34282	36578	33068	28654	32323	29577
Combustible fuels	27542	28748	23277	18367	21187	16483
Hydro	19	21	17	17	13	15
Nuclear	..	..	..	..	..	..
Other	6721	7809	9774	10270	11123	13079
Own use in electricity, CHP and heat plants	1921	1989	1680	1500	1604	1369
Net production	34462	36873	33549	29201	33156	30814
Imports	11208	10599	11694	15920	11459	12702
Exports	10874	11734	10374	10706	10377	9847
Losses	2365	2624	2201	2150	1931	1974
Consumption	32409	33018	32584	32073	32151	31569
Energy industries own use	964	960	925	892	938	944
By industry and construction	8398	8514	8648	8543	8345	8174
By transport	395	404	397	385	386	385
By households and other cons.	22652	23140	22614	22253	22482	22066
Net installed capacity	**Thousand kilowatts**					
Total main activity and autoproducer	13393	13438	13587	14075	13810	13656
Combustible fuels	9897	9620	9609	9500	8410	8152
Hydro	9	9	9	9	9	9
Nuclear	..	..	..	..	..	..
Other	3487	3809	3969	4566	5391	5495
Main activity	12741	12795	12938	13043	12665	12476
Combustible fuels	9250	8984	8977	8870	7836	7579
Hydro	9	9	9	9	9	9
Nuclear	..	..	..	..	..	..
Other	3482	3802	3952	4164	4820	4888
Combustible fuel input	**Terajoules**					
Hard Coal	163446	157822	130137	101277	130091	102110
Biogas	3266	3414	3269	3516	3706	4223
Gas-diesel oil	2666	1591	1290	1118	946	430
Fuel oil	9090	6060	2626	2141	2384	1495
Liquefied petroleum gas	..	47	47	47	47	47
Refinery gas	1436	1535	1188	1337	1337	1584
Natural gas	70767	84297	62624	47273	39542	23296
Fuelwood	*23917	*40152	*38156	*40969	*42200	*41155
Vegetal waste	0	0	0	0	0	0
Municipal waste	33880	32654	32728	32282	32301	33473
Others	0	0	0	0	0	0
Total input	308467	327572	272065	229960	252553	207813
Total production	106700	111694	91523	73116	83182	66575
Estimated efficiency (% of production to input)	35	34	34	32	33	32

Statistics on electricity

Djibouti

Item	2009	2010	2011	2012	2013	2014
Production, trade and consumption	**Million kilowatt-hours**					
Total main activity and autoproducer	349	*379	*387	393	402	402
Combustible fuels	349	*379	*387	393	402	402
Hydro	..	..	..	..	..	..
Nuclear	..	..	..	..	..	..
Other	..	..	..	..	..	..
Main activity	349	*379	*387	393	402	402
Combustible fuels	349	*379	*387	393	402	402
Hydro	..	..	..	..	..	..
Nuclear	..	..	..	..	..	..
Other	..	..	..	..	..	..
Own use in electricity, CHP and heat plants	6	*6	*7	6	6	6
Net production	343	*373	*381	387	396	396
Imports	..	..	..	..	..	..
Exports	..	..	..	..	..	..
Losses	75	84	90	*90	*90	*91
Consumption	*267	*289	*290	*297	*306	*305
Energy industries own use	..	..	..	..	..	..
By industry and construction	..	..	..	..	..	..
By transport	..	..	..	..	..	..
By households and other cons.	*267	*289	*290	*297	*306	*305
Net installed capacity	**Thousand kilowatts**					
Total main activity and autoproducer	130	130	130	130	130	130
Combustible fuels	130	130	130	130	130	130
Hydro	..	..	..	..	..	..
Nuclear	..	..	..	..	..	..
Other	..	..	..	..	..	..
Main activity	130	130	130	130	130	130
Combustible fuels	130	130	130	130	130	130
Hydro	..	..	..	..	..	..
Nuclear	..	..	..	..	..	..
Other	..	..	..	..	..	..
Combustible fuel input	**Terajoules**					
Gas-diesel oil	*1406	*1084	*1170	*1234	*1294	*1320
Fuel oil	*2222	*2828	*2828	*2828	*2969	*3030
Total input	*3628	*3912	*3998	*4062	*4264	*4350
Total production	1256	*1364	*1393	1415	1447	1447
Estimated efficiency (% of production to input)	35	35	35	35	34	33

Statistics on electricity

Dominica

Item	2009	2010	2011	2012	2013	2014
Production, trade and consumption	Million kilowatt-hours					
Total main activity and autoproducer	93	99	100	102	101	102
Combustible fuels	70	76	65	75	64	72
Hydro	23	23	36	27	37	31
Nuclear	..	..	..	..	..	..
Other	..	..	..	..	..	..
Main activity	93	99	100	102	101	102
Combustible fuels	70	76	65	75	64	72
Hydro	23	23	36	27	37	31
Nuclear	..	..	..	..	..	..
Other	..	..	..	..	..	..
Own use in electricity, CHP and heat plants	3	3	3	4	3	3
Net production	90	96	97	98	97	99
Imports	..	..	..	..	..	..
Exports	..	..	..	..	..	..
Losses	9	9	8	8	8	8
Consumption	80	87	89	90	89	91
Energy industries own use	..	..	..	..	..	..
By industry and construction	8	8	8	8	8	9
By transport	..	..	..	..	..	..
By households and other cons.	72	79	81	82	81	82
Net installed capacity	Thousand kilowatts					
Total main activity and autoproducer	24	27	27	27	27	27
Combustible fuels	19	20	20	20	20	20
Hydro	5	7	7	7	7	7
Nuclear	..	..	..	..	..	..
Other	..	..	..	..	..	..
Main activity	24	27	27	27	27	27
Combustible fuels	19	20	20	20	20	20
Hydro	5	7	7	7	7	7
Nuclear	..	..	..	..	..	..
Other	..	..	..	..	..	..
Combustible fuel input	Terajoules					
Gas-diesel oil	679	753	636	740	624	684
Total input	679	753	636	740	624	684
Total production	251	274	233	269	230	258
Estimated efficiency (% of production to input)	37	36	37	36	37	38

Statistics on electricity

Dominican Republic

Item	2009	2010	2011	2012	2013	2014
Production, trade and consumption	**Million kilowatt-hours**					
Total main activity and autoproducer	15060	16116	16701	17746	18799	17962
Combustible fuels	13606	14700	15174	15867	16685	16452
Hydro	1453	1416	1513	1783	1872	1267
Nuclear	..	..	..	..	..	..
Other	0	0	14	96	243	243
Main activity	12338	13257	13782	14690	15610	14992
Combustible fuels	10885	11841	12254	12812	13495	13482
Hydro	1453	1416	1513	1783	1872	1267
Nuclear	..	..	..	..	..	..
Other	0	0	14	96	243	243
Own use in electricity, CHP and heat plants	420	415	492	510	858	592
Net production	14639	15701	16209	17236	17942	17370
Imports	..	..	..	..	..	..
Exports	..	..	..	..	..	..
Losses	1788	1926	1993	2127	2213	2160
Consumption	12845	13774	14215	15109	15729	15212
Energy industries own use	21	23	23	24	25	23
By industry and construction	5046	5419	5595	5772	6497	5374
By transport	25	27	30	31	46	52
By households and other cons.	7752	8307	8567	9282	9160	9763
Net installed capacity	**Thousand kilowatts**					
Total main activity and autoproducer	3381	3378	3440	3678	4120	4042
Combustible fuels	2858	2855	2883	2986	3421	3328
Hydro	523	523	523	605	606	616
Nuclear	..	..	..	..	..	..
Other	0	0	33	87	93	98
Main activity	3207	3204	3265	3501	3713	3616
Combustible fuels	2683	2680	2709	2811	3021	2916
Hydro	523	523	523	605	606	616
Nuclear	..	..	..	..	..	..
Other	0	0	33	85	85	84
Combustible fuel input	**Terajoules**					
Hard Coal	20243	18991	20493	22611	22250	25468
Motor gasoline	631	662	676	709	720	688
Gas-diesel oil	23793	19597	26909	25398	24013	20855
Fuel oil	57736	58721	53211	51560	58520	57204
Natural gas	17918	28386	29937	35004	36487	35778
Bagasse	865	678	644	710	666	666
Total input	121187	127036	131871	135991	142656	140659
Total production	48983	52920	54625	57121	60065	59227
Estimated efficiency (% of production to input)	40	42	41	42	42	42

Ecuador

Item	2009	2010	2011	2012	2013	2014
Production, trade and consumption	Million kilowatt-hours					
Total main activity and autoproducer	18265	19510	20544	22848	23260	24307
Combustible fuels	9036	10870	9408	10608	12161	12753
Hydro	9225	8636	11133	12238	11039	11458
Nuclear	..	..	..	..	..	..
Other	3	3	3	3	60	96
Main activity	15219	16256	17103	18982	19369	20477
Combustible fuels	6479	8073	6472	7252	8783	9445
Hydro	8737	8179	10628	11727	10525	10936
Nuclear	..	..	..	..	..	..
Other	3	3	3	3	60	96
Own use in electricity, CHP and heat plants	360	328	352	385	420	515
Net production	17904	19182	20192	22463	22840	23792
Imports	1121	873	1295	238	662	837
Exports	21	10	14	12	29	47
Losses	3371	3260	3258	3180	3010	3140
Consumption	15448	16792	18185	19492	20430	21491
Energy industries own use	..	..	..	..	..	..
By industry and construction	6368	7122	7724	8320	8360	8497
By transport	10	10	10	10	10	10
By households and other cons.	9070	9660	10451	11163	12059	12984
Net installed capacity	Thousand kilowatts					
Total main activity and autoproducer	4396	4757	4796	5063	5103	5299
Combustible fuels	2361	2540	2586	2824	2843	3011
Hydro	2032	2215	2207	2237	2237	2241
Nuclear	..	..	..	..	..	..
Other	2	2	2	2	23	48
Main activity	3750	4046	4083	4323	4308	4434
Combustible fuels	1798	1910	1947	2160	2125	2220
Hydro	1949	2133	2134	2161	2161	2166
Nuclear	..	..	..	..	..	..
Other	2	2	2	2	23	48
Combustible fuel input	Terajoules					
Crude oil	7835	8315	8628	9226	10387	10590
Motor gasoline	1254	1843	1852	13	341	0
Gas-diesel oil	29763	45147	24675	19932	25292	26578
Fuel oil	28904	30241	34156	40164	44160	47372
Liquefied petroleum gas	1141	1166	1063	947	882	1199
Other oil products	..	0	1769	0	0	0
Natural gas	19170	21223	18754	24606	27393	28227
Bagasse	6922	7319	8538	9005	8772	10692
Total input	94988	115255	99436	103893	117228	124658
Total production	32530	39132	33867	38187	43780	45911
Estimated efficiency (% of production to input)	34	34	34	37	37	37

Statistics on electricity

Egypt

Item	2009	2010	2011	2012	2013	2014
Production, trade and consumption	**Million kilowatt-hours**					
Total main activity and autoproducer	142690	150486	161162	164628	168050	177249
Combustible fuels	128694	135736	146224	150010	153252	161708
Hydro	12863	13046	12934	13121	13352	13979
Nuclear	..	..	..	..	..	..
Other	1133	1704	2004	1497	1446	1562
Main activity	139000	146796	157406	159223	162548	171747
Combustible fuels	125004	132046	142468	144605	147750	156206
Hydro	12863	13046	12934	13121	13352	13979
Nuclear	..	..	..	..	..	..
Other	1133	1704	2004	1497	1446	1562
Own use in electricity, CHP and heat plants	4545	4944	5226	5405	5502	11150
Net production	138145	145542	155936	159223	162548	166099
Imports	183	152	102	77	61	81
Exports	1118	1595	1679	474	460	470
Losses	15552	14918	16342	18305	24224	19154
Consumption	120180	126934	135838	140257	137591	146556
Energy industries own use	..	..	..	..	..	..
By industry and construction	38916	40702	42098	39887	37320	41678
By transport	..	..	..	506	441	528
By households and other cons.	81264	86232	93740	99864	99830	104350
Net installed capacity	**Thousand kilowatts**					
Total main activity and autoproducer	25409	27732	29997	31487	32702	35907
Combustible fuels	22119	24245	26275	28000	29215	32420
Hydro	2800	2800	3035	2800	2800	2800
Nuclear	..	..	..	..	..	..
Other	490	687	687	687	687	687
Main activity	24726	27049	29310	30800	32015	35220
Combustible fuels	21436	23562	25588	27313	28528	31733
Hydro	2800	2800	3035	2800	2800	2800
Nuclear	..	..	..	..	..	..
Other	490	687	687	687	687	687
Combustible fuel input	**Terajoules**					
Gas-diesel oil	7525	3698	2580	4257	*4257	3569
Fuel oil	226240	210242	184507	265913	315686	224058
Natural gas	931722	910127	1061998	1068359	1009816	1082510
Total input	1165487	1124067	1249085	1338529	1329759	1310137
Total production	463298	488650	526406	540036	551707	582149
Estimated efficiency (% of production to input)	40	43	42	40	41	44

Statistics on electricity

El Salvador

Item	2009	2010	2011	2012	2013	2014
Production, trade and consumption	Million kilowatt-hours					
Total main activity and autoproducer	5792	5984	6083	6237	6272	6223
Combustible fuels	2763	2375	2538	2855	2922	2947
Hydro	1505	2084	2011	1847	1790	1718
Nuclear	..	..	..	..	..	..
Other	1524	1525	1534	1535	1560	1558
Main activity	5153	5369	5490	5541	5612	5569
Combustible fuels	2124	1760	1945	2159	2262	2293
Hydro	1505	2084	2011	1847	1790	1718
Nuclear	..	..	..	..	..	..
Other	1524	1525	1534	1535	1560	1558
Own use in electricity, CHP and heat plants	275	271	300	309	363	344
Net production	5517	5713	5783	5928	5909	5879
Imports	208	174	216	163	283	589
Exports	79	89	102	78	91	208
Losses	711	769	728	610	455	705
Consumption	4933	5024	5169	5383	5647	5524
Energy industries own use	..	..	..	..	..	..
By industry and construction	2204	2231	2287	1741	2197	2231
By transport	..	..	..	..	..	..
By households and other cons.	2729	2793	2882	3642	3450	3293
Net installed capacity	Thousand kilowatts					
Total main activity and autoproducer	1471	1461	1477	1478	1537	1563
Combustible fuels	795	785	801	801	860	886
Hydro	472	472	472	473	473	473
Nuclear	..	..	..	..	..	..
Other	204	204	204	204	204	204
Main activity	1271	1284	1284	1285	1344	1370
Combustible fuels	615	608	608	608	667	693
Hydro	452	472	472	473	473	473
Nuclear	..	..	..	..	..	..
Other	204	204	204	204	204	204
Combustible fuel input	Terajoules					
Biogas	..	..	..	..	526	409
Gas-diesel oil	129	86	86	215	129	43
Fuel oil	20644	17210	18463	18988	21008	21331
Bagasse	5807	5878	5343	12347	7495	7640
Total input	26580	23174	23892	31550	29158	29423
Total production	9947	8550	9137	10278	10519	10609
Estimated efficiency (% of production to input)	37	37	38	33	36	36

Equatorial Guinea

Item	2009	2010	2011	2012	2013	2014
Production, trade and consumption	**Million kilowatt-hours**					
Total main activity and autoproducer	*122	*122	*281	*439	*439	*439
Combustible fuels	*83	*83	*242	*242	*242	*242
Hydro	*39	*39	*39	*197	*197	*197
Nuclear	..	..	..	..	..	..
Other	..	..	..	..	..	..
Main activity	*122	*122	*281	*439	*439	*439
Combustible fuels	*83	*83	*242	*242	*242	*242
Hydro	*39	*39	*39	*197	*197	*197
Nuclear	..	..	..	..	..	..
Other	..	..	..	..	..	..
Own use in electricity, CHP and heat plants	*12	*12	*28	*44	*44	*44
Net production	*110	*110	*253	*395	*395	*395
Imports	..	..	..	..	..	..
Exports	..	..	..	..	..	..
Losses	..	..	..	..	..	..
Consumption	*110	*110	*253	*395	*395	*395
Energy industries own use	*10	*10	*23	*35	*35	*35
By industry and construction	*51	*51	*118	*184	*184	*184
By transport	..	..	..	..	..	..
By households and other cons.	*49	*49	*112	*176	*176	*176
Net installed capacity	**Thousand kilowatts**					
Total main activity and autoproducer	*93	*93	*214	*334	*334	*334
Combustible fuels	*63	*63	*184	*184	*184	*184
Hydro	*30	*30	*30	*150	*150	*150
Nuclear	..	..	..	..	..	..
Other	..	..	..	..	..	..
Main activity	*93	*93	*214	*334	*334	*334
Combustible fuels	*63	*63	*184	*184	*184	*184
Hydro	*30	*30	*30	*150	*150	*150
Nuclear	..	..	..	..	..	..
Other	..	..	..	..	..	..
Combustible fuel input	**Terajoules**					
Gas-diesel oil	*86	*86	*86	*86	*86	*86
Natural gas	*907	*907	*2401	*2401	2401	*2401
Total input	*993	*993	*2487	*2487	2487	*2487
Total production	*299	*299	*871	*871	*871	*871
Estimated efficiency (% of production to input)	30	30	35	35	35	35

Statistics on electricity

Eritrea

Item	2009	2010	2011	2012	2013	2014
Production, trade and consumption	**Million kilowatt-hours**					
Total main activity and autoproducer	295	311	337	359	363	370
Combustible fuels	293	309	335	357	361	368
Hydro	..	..	..	..	..	..
Nuclear	..	..	..	..	..	..
Other	2	2	2	2	2	2
Main activity	285	300	326	348	352	359
Combustible fuels	283	298	324	346	350	357
Hydro	..	..	..	..	..	..
Nuclear	..	..	..	..	..	..
Other	2	2	2	2	2	2
Own use in electricity, CHP and heat plants	17	15	16	17	17	17
Net production	278	296	321	342	346	353
Imports	..	..	..	..	..	..
Exports	..	..	..	..	..	..
Losses	35	39	50	57	51	50
Consumption	242	256	271	285	295	303
Energy industries own use	..	..	..	..	..	..
By industry and construction	65	64	70	75	81	84
By transport	..	..	..	..	..	..
By households and other cons.	177	192	201	210	214	219
Net installed capacity	**Thousand kilowatts**					
Total main activity and autoproducer	*140	*140	*140	*140	*140	*140
Combustible fuels	*139	*139	*139	*139	*139	*139
Hydro	..	..	..	..	..	..
Nuclear	..	..	..	..	..	..
Other	*1	*1	*1	*1	*1	*1
Main activity	*134	*134	*134	*134	*134	*134
Combustible fuels	*133	*133	*133	*133	*133	*133
Hydro	..	..	..	..	..	..
Nuclear	..	..	..	..	..	..
Other	*1	*1	*1	*1	*1	*1
Combustible fuel input	**Terajoules**					
Gas-diesel oil	1118	1247	1333	1419	1419	1462
Fuel oil	2141	2262	2464	2626	2666	2707
Total input	3259	3509	3797	4045	4085	4169
Total production	1055	1112	1206	1285	1300	1325
Estimated efficiency (% of production to input)	32	32	32	32	32	32

Statistics on electricity

Estonia

Item	2009	2010	2011	2012	2013	2014
Production, trade and consumption	**Million kilowatt-hours**					
Total main activity and autoproducer	8779	12964	12893	11967	13275	12446
Combustible fuels	8552	12660	12495	11491	12720	11815
Hydro	32	27	30	42	26	27
Nuclear	..	..	..	..	..	..
Other	195	277	368	434	529	604
Main activity	8683	12857	12788	11869	13184	12389
Combustible fuels	8457	12554	12390	11394	12634	11773
Hydro	32	27	30	42	26	26
Nuclear	..	..	..	..	..	..
Other	194	276	368	433	524	590
Own use in electricity, CHP and heat plants	895	1232	1226	1440	1452	1433
Net production	7884	11732	11667	10527	11823	11013
Imports	3025	1100	1690	2710	2712	3730
Exports	2943	4354	5252	4950	6300	6484
Losses	886	1047	949	879	903	842
Consumption	7080	7431	7156	7408	7332	7417
Energy industries own use	430	523	529	430	512	511
By industry and construction	1941	2095	2046	2187	2156	2117
By transport	90	89	80	79	63	50
By households and other cons.	4619	4724	4501	4712	4601	4739
Net installed capacity	**Thousand kilowatts**					
Total main activity and autoproducer	2665	2751	2825	2923	2910	3096
Combustible fuels	2554	2637	2640	2649	2654	2750
Hydro	7	6	5	8	8	5
Nuclear	..	..	..	..	..	..
Other	104	108	180	266	248	341
Main activity	2647	2729	2806	2904	2892	3072
Combustible fuels	2537	2616	2621	2630	2637	2733
Hydro	7	6	5	8	8	5
Nuclear	..	..	..	..	..	..
Other	103	107	180	266	247	334
Combustible fuel input	**Terajoules**					
Hard Coal	0	0	0	0	54	54
Peat	919	1366	1056	1230	920	690
Biogas	65	111	93	92	200	191
Oil shale	87523	120399	124378	111401	137372	135538
Gas-diesel oil	43	0	0	0	0	0
Fuel oil	444	485	364	687	929	485
Natural gas	4034	3771	3026	1411	779	564
Gasworks gas	5181	5943	5415	4866	3899	5228
Fuelwood	*3368	*7338	*7404	*9978	*9108	*9940
Municipal waste	..	..	..	..	1336	1455
Total input	101577	139413	141735	129665	154597	154145
Total production	30787	45576	44982	41368	45792	42534
Estimated efficiency (% of production to input)	30	33	32	32	30	28

Statistics on electricity

Ethiopia

Item	2009	2010	2011	2012	2013	2014
Production, trade and consumption	Million kilowatt-hours					
Total main activity and autoproducer	3992	4980	6336	7608	8719	9615
Combustible fuels	444	31	37	12	8	9
Hydro	3524	4931	6262	7388	8338	9195
Nuclear	..	..	..	..	..	..
Other	24	18	37	208	373	411
Main activity	3992	4980	6336	7608	8719	9615
Combustible fuels	444	31	37	12	8	9
Hydro	3524	4931	6262	7388	8338	9195
Nuclear	..	..	..	..	..	..
Other	24	18	37	208	373	411
Own use in electricity, CHP and heat plants	340	396	367	29	23	259
Net production	3652	4584	5969	7579	8696	9356
Imports	..	..	..	..	..	..
Exports	..	0	332	563	954	1052
Losses	399	747	1267	1740	1655	1775
Consumption	3253	3837	4370	5276	6087	6529
Energy industries own use	..	..	..	..	..	..
By industry and construction	1223	1395	1590	1943	2032	2241
By transport	..	..	..	..	..	..
By households and other cons.	2030	2442	2780	3333	4055	4288
Net installed capacity	Thousand kilowatts					
Total main activity and autoproducer	2214	2155	2310	2310	2311	2311
Combustible fuels	188	129	189	189	190	190
Hydro	1848	1848	1943	1943	1943	1943
Nuclear	..	..	..	..	..	..
Other	178	178	178	178	178	178
Main activity	2214	2155	2310	2310	2311	2311
Combustible fuels	188	129	189	189	190	190
Hydro	1848	1848	1943	1943	1943	1943
Nuclear	..	..	..	..	..	..
Other	178	178	178	178	178	178
Combustible fuel input	Terajoules					
Gas-diesel oil	6579	473	559	172	129	129
Total input	6579	473	559	172	129	129
Total production	1598	112	133	43	29	32
Estimated efficiency (% of production to input)	24	24	24	25	22	25

Statistics on electricity

Faeroe Islands

Item	2009	2010	2011	2012	2013	2014
Production, trade and consumption	**Million kilowatt-hours**					
Total main activity and autoproducer	276	280	274	292	293	305
Combustible fuels	168	199	167	181	180	150
Hydro	92	67	93	100	91	121
Nuclear	..	..	..	..	..	..
Other	15	14	15	11	22	34
Main activity	276	280	274	292	293	305
Combustible fuels	168	199	167	181	180	150
Hydro	92	67	93	100	91	121
Nuclear	..	..	..	..	..	..
Other	15	14	15	11	22	34
Own use in electricity, CHP and heat plants	7	7	7	7	8	7
Net production	269	273	267	285	285	298
Imports	..	..	..	..	..	..
Exports	..	..	..	..	..	..
Losses	0	*18	*12	*23	*10	*15
Consumption	269	255	255	261	275	284
Energy industries own use	..	..	..	..	..	..
By industry and construction	57	49	45	52	60	76
By transport	..	..	..	..	..	..
By households and other cons.	212	206	210	209	214	208
Net installed capacity	**Thousand kilowatts**					
Total main activity and autoproducer	103	103	103	111	*115	*124
Combustible fuels	65	65	65	65	*65	*68
Hydro	32	32	32	*39	*39	*39
Nuclear	..	..	..	..	..	..
Other	*7	*7	*7	*7	*11	*16
Main activity	103	103	103	111	*115	*124
Combustible fuels	65	65	65	65	*65	*68
Hydro	32	32	32	*39	*39	*39
Nuclear	..	..	..	..	..	..
Other	*7	*7	*7	*7	*11	*16
Combustible fuel input	**Terajoules**					
Gas-diesel oil	128	69	43	28	44	50
Fuel oil	1394	1634	1477	1485	1506	1248
Total input	1522	1703	1520	1512	1549	1298
Total production	605	718	601	652	648	541
Estimated efficiency (% of production to input)	40	42	40	43	42	42

Falkland Is. (Malvinas)

Item	2009	2010	2011	2012	2013	2014
Production, trade and consumption	Million kilowatt-hours					
Total main activity and autoproducer	*18	*18	*18	*18	*18	*18
Combustible fuels	*17	*12	*12	*12	*12	*12
Hydro	..	..	..	..	..	..
Nuclear	..	..	..	..	..	..
Other	0	*6	*6	*6	*6	*6
Main activity	*16	*16	*16	*16	*16	*16
Combustible fuels	*15	*10	*10	*10	*10	*10
Hydro	..	..	..	..	..	..
Nuclear	..	..	..	..	..	..
Other	0	*6	*6	*6	*6	*6
Own use in electricity, CHP and heat plants	0	0	0	0	0	0
Net production	*18	*18	*18	*18	*18	*18
Imports	..	..	..	..	..	..
Exports	..	..	..	..	..	..
Losses	..	..	..	..	..	..
Consumption	*18	*18	*18	*18	*18	*18
Energy industries own use	..	..	..	..	..	..
By industry and construction	*7	*7	*7	*7	*7	*7
By transport	..	..	..	..	..	..
By households and other cons.	*10	*10	*10	*10	*10	*10
Net installed capacity	Thousand kilowatts					
Total main activity and autoproducer	*9	*12	*12	*12	*12	*12
Combustible fuels	*9	*9	*9	*9	*9	*9
Hydro	..	..	..	..	..	..
Nuclear	..	..	..	..	..	..
Other	0	*3	*3	*3	*3	*3
Main activity	*7	*10	*10	*10	*10	*10
Combustible fuels	*7	*7	*7	*7	*7	*7
Hydro	..	..	..	..	..	..
Nuclear	..	..	..	..	..	..
Other	0	*3	*3	*3	*3	*3
Combustible fuel input	Terajoules					
Gas-diesel oil	*172	*112	*112	*112	*112	*112
Total input	*172	*112	*112	*112	*112	*112
Total production	*62	*43	*43	*43	*43	*43
Estimated efficiency (% of production to input)	36	38	38	39	39	39

Statistics on electricity

Fiji

Item	2009	2010	2011	2012	2013	2014
Production, trade and consumption	**Million kilowatt-hours**					
Total main activity and autoproducer	777	835	801	803	858	859
Combustible fuels	310	415	340	271	325	454
Hydro	460	414	456	525	527	401
Nuclear	..	..	..	..	..	..
Other	7	6	5	7	5	4
Main activity	734	783	757	755	811	810
Combustible fuels	266	363	295	223	278	404
Hydro	460	414	456	525	527	401
Nuclear	..	..	..	..	..	..
Other	7	6	5	7	5	4
Own use in electricity, CHP and heat plants	9	9	9	8	9	23
Net production	768	826	792	795	848	836
Imports	..	..	..	..	..	..
Exports	..	..	..	..	..	..
Losses	*58	*62	*42	*63	*68	*70
Consumption	715	764	767	732	780	783
Energy industries own use	..	..	..	..	..	..
By industry and construction	172	190	198	192	202	200
By transport	..	..	..	..	..	..
By households and other cons.	543	574	569	540	578	583
Net installed capacity	**Thousand kilowatts**					
Total main activity and autoproducer	*215	244	*260	285	309	309
Combustible fuels	*94	105	*120	*140	164	164
Hydro	*111	129	*129	135	135	135
Nuclear	..	..	..	..	..	..
Other	10	10	10	10	10	10
Main activity	*171	194	*210	235	259	259
Combustible fuels	*50	55	*70	*90	114	114
Hydro	*111	129	*129	135	135	135
Nuclear	..	..	..	..	..	..
Other	10	10	10	10	10	10
Combustible fuel input	**Terajoules**					
Hard coal briquettes	0	0	0	0	0	0
Gas-diesel oil	1806	2623	2322	1376	1291	2560
Fuel oil	1010	1131	727	1172	1616	1535
Total input	2816	3754	3049	2548	2907	4095
Total production	1116	1494	1223	977	1169	1634
Estimated efficiency (% of production to input)	40	40	40	38	40	40

Statistics on electricity

Finland

Item	2009	2010	2011	2012	2013	2014
Production, trade and consumption	Million kilowatt-hours					
Total main activity and autoproducer	72062	80679	73506	70416	71263	68093
Combustible fuels	35266	44345	37083	29746	33744	29711
Hydro	12686	12922	12445	16859	12838	13397
Nuclear	23526	22800	23187	22987	23606	23580
Other	584	612	791	824	1075	1405
Main activity	62092	69594	62852	60161	61141	58000
Combustible fuels	26388	34320	27573	20955	24784	20735
Hydro	11790	12062	11496	15615	11859	12470
Nuclear	23526	22800	23187	22987	23606	23580
Other	388	412	596	604	892	1215
Own use in electricity, CHP and heat plants	2854	3459	3093	2709	2910	2637
Net production	69208	77220	70413	67707	68353	65456
Imports	15460	15719	17656	19089	17591	21622
Exports	3375	5218	3804	1645	1876	3655
Losses	2774	2765	2698	2914	2609	2772
Consumption	78404	84790	81408	82027	81271	80408
Energy industries own use	1338	1291	1301	1310	1331	1271
By industry and construction	36164	40360	39241	38165	38692	38170
By transport	716	740	730	737	733	722
By households and other cons.	40186	42399	40136	41815	40515	40245
Net installed capacity	Thousand kilowatts					
Total main activity and autoproducer	15335	15536	15677	15757	16653	16245
Combustible fuels	9321	9461	9559	9564	10221	9607
Hydro	3145	3155	3196	3196	3224	3248
Nuclear	2716	2716	2716	2732	2752	2752
Other	153	204	206	265	456	638
Main activity	13267	13467	13639	13749	14610	14200
Combustible fuels	7454	7594	7723	7731	8373	7759
Hydro	2950	2960	3001	3029	3038	3062
Nuclear	2716	2716	2716	2732	2752	2752
Other	147	197	199	257	447	627
Combustible fuel input	Terajoules					
Hard Coal	110331	142647	99982	80119	110621	83563
Peat	55886	76374	66826	49985	43441	43898
Biogas	225	727	1133	953	2387	2800
Fuel oil	5373	4888	3636	3111	2707	2707
Natural gas	85887	97970	82347	63266	59170	46749
Blast furnace gas	4542	5818	5642	4866	4597	5597
Fuelwood	*99741	*118763	*123943	*130689	*143253	*29222
Municipal waste	6588	6680	6971	9365	12215	13468
Industrial waste	806	697	812	935	1071	1083
Black liquor	*61265	*89146	*85351	*84307	*89097	*78683
Others	2180	1431	1396	2088	2609	2339
Total input	432824	545141	478040	*429684	471168	310110
Total production	126958	159642	133499	107086	121478	106960
Estimated efficiency (% of production to input)	29	29	28	25	26	34

Statistics on electricity

France

Item	2009	2010	2011	2012	2013	2014
Production, trade and consumption	**Million kilowatt-hours**					
Total main activity and autoproducer	535635	569097	561448	565704	572308	562776
Combustible fuels	55396	62010	54594	56722	50877	33484
Hydro	61969	67525	49865	63594	75867	68626
Nuclear	409736	428521	442383	425406	423685	436474
Other	8534	11041	14606	19982	21879	24192
Main activity	522440	556404	545596	543822	553522	544147
Combustible fuels	43188	50950	41339	39502	37652	20865
Hydro	61380	66884	49389	62940	75124	67826
Nuclear	409736	428521	442383	425406	423685	436474
Other	8136	10049	12485	15974	17061	18982
Own use in electricity, CHP and heat plants	23638	24985	25017	24522	23803	23360
Net production	511997	544112	536431	541182	548505	539416
Imports	18517	19475	9501	12213	11687	7873
Exports	44451	50188	65914	56734	60148	75063
Losses	34878	35414	32412	37702	37579	35384
Consumption	454430	478359	449894	460938	464169	439561
Energy industries own use	36475	34270	32328	26845	23459	24236
By industry and construction	111722	117444	117891	114319	111440	111382
By transport	12518	12533	12398	12407	12780	12473
By households and other cons.	293715	314112	287277	307367	316490	291470
Net installed capacity	**Thousand kilowatts**					
Total main activity and autoproducer	119038	124551	127256	129254	128432	129069
Combustible fuels	25624	28824	27792	27764	25576	24411
Hydro	25185	25401	25347	25366	25360	25294
Nuclear	63130	63130	63130	63130	63130	63130
Other	5099	7196	10987	12994	14366	16234
Main activity	113348	117468	119210	120208	120031	120835
Combustible fuels	20606	23179	23654	23491	22162	21655
Hydro	24929	25198	25000	25000	25153	25084
Nuclear	63130	63130	63130	63130	63130	63130
Other	4683	5961	7426	8587	9586	10966
Combustible fuel input	**Terajoules**					
Hard Coal	219206	206383	143077	193664	219590	96738
Biogas	9767	11180	11757	12769	14214	13996
Fuel oil	55025	55146	26422	26583	15029	10221
Refinery gas	10742	18612	19602	19503	13662	9059
Other oil products	16241	12221	13829	12784	6995	7276
Natural gas	265241	349935	366062	209649	166357	114927
Blast furnace gas	17363	21893	19676	21728	23371	23211
Fuelwood	*26002	*35309	*17489	*15250	*18572	*21018
Municipal waste	76044	77584	77218	82366	78660	78402
Black liquor	*11339	*16285	*34669	*24148	*20496	*23852
Others	14232	13577	9077	10719	12298	10627
Total input	721201	818125	738878	629163	589243	409327
Total production	199426	223236	196538	204199	183157	120542
Estimated efficiency (% of production to input)	28	27	27	32	31	29

Statistics on electricity

French Guiana

Item	2009	2010	2011	2012	2013	2014
Production, trade and consumption	**Million kilowatt-hours**					
Total main activity and autoproducer	*845	872	873	874	883	910
Combustible fuels	*335	402	373	266	349	359
Hydro	*510	469	469	556	490	500
Nuclear	..	..	..	..	..	..
Other	..	1	31	52	44	51
Main activity	*845	872	873	874	883	910
Combustible fuels	*335	402	373	266	349	359
Hydro	*510	469	469	556	490	500
Nuclear	..	..	..	..	..	..
Other	..	1	31	52	44	51
Own use in electricity, CHP and heat plants	*10	12	10	15	17	20
Net production	*835	860	863	859	866	890
Imports	..	..	..	..	..	..
Exports	..	..	..	..	..	..
Losses	*70	*86	*86	*94	*97	*115
Consumption	775	743	749	765	769	775
Energy industries own use	..	..	..	..	..	..
By industry and construction	43	21	24	23	46	43
By transport	..	..	..	..	..	..
By households and other cons.	732	722	725	742	723	732
Net installed capacity	**Thousand kilowatts**					
Total main activity and autoproducer	243	243	247	281	292	292
Combustible fuels	129	129	129	129	140	140
Hydro	114	114	118	118	119	119
Nuclear	..	..	..	..	..	..
Other	..	0	0	34	33	33
Main activity	243	243	247	281	292	292
Combustible fuels	129	129	129	129	140	140
Hydro	114	114	118	118	119	119
Nuclear	..	..	..	..	..	..
Other	..	0	0	34	33	33
Combustible fuel input	**Terajoules**					
Gas-diesel oil	*1290	*1591	*1591	*1376	*1591	*1677
Fuel oil	*1858	*1919	*1899	*1212	*1535	*1616
Total input	*3148	*3510	*3490	*2588	*3126	*3293
Total production	*1206	1447	1343	958	1256	1292
Estimated efficiency (% of production to input)	38	41	38	37	40	39

French Polynesia

Item	2009	2010	2011	2012	2013	2014
Production, trade and consumption	**Million kilowatt-hours**					
Total main activity and autoproducer	*736	*817	*761	*768	*744	*721
Combustible fuels	*521	*538	*533	*553	*542	*492
Hydro	210	276	*225	*209	*195	*219
Nuclear	..	..	..	..	..	..
Other	4	*3	*3	*6	*7	*10
Main activity	*736	*817	*761	*768	744	*721
Combustible fuels	*521	*538	*533	*553	*542	*492
Hydro	210	276	*225	*209	*195	*219
Nuclear	..	..	..	..	..	..
Other	4	*3	*3	*6	*7	*10
Own use in electricity, CHP and heat plants	29	32	*30	*33	*34	*35
Net production	*706	*785	*731	*735	*710	*686
Imports	..	..	..	..	..	..
Exports	..	..	..	..	..	..
Losses	*44	*49	*45	*46	*46	*44
Consumption	*670	*736	*685	689	665	*642
Energy industries own use	..	..	..	..	..	..
By industry and construction	..	..	..	..	..	..
By transport	..	..	..	..	..	..
By households and other cons.	*670	*736	*685	689	665	*642
Net installed capacity	**Thousand kilowatts**					
Total main activity and autoproducer	228	*227	*227	*227	*227	*227
Combustible fuels	177	*177	*177	*177	*177	*177
Hydro	47	*47	*47	*47	*48	*48
Nuclear	..	..	..	..	..	..
Other	3	*3	*3	*3	*3	*3
Main activity	228	*227	*227	*227	*227	*227
Combustible fuels	177	*177	*177	*177	*177	*177
Hydro	47	*47	*47	*47	*48	*48
Nuclear	..	..	..	..	..	..
Other	3	*3	*3	*3	*3	*3
Combustible fuel input	**Terajoules**					
Gas-diesel oil	*3223	*3239	*3008	*2869	*2937	*2884
Fuel oil	*1398	*1405	*1405	*1405	*1420	1364
Total input	*4621	*4644	*4413	*4274	*4357	*4248
Total production	*1877	*1936	*1918	*1989	*1951	*1772
Estimated efficiency (% of production to input)	41	42	43	47	45	42

Statistics on electricity

Gabon

Item	2009	2010	2011	2012	2013	2014
Production, trade and consumption	Million kilowatt-hours					
Total main activity and autoproducer	1835	1932	2033	2151	2264	2367
Combustible fuels	891	1025	1193	1240	1330	1570
Hydro	944	907	840	908	932	795
Nuclear	..	..	..	..	..	..
Other	..	..	..	3	2	2
Main activity	1652	1752	1838	1966	2071	2173
Combustible fuels	708	845	998	1058	1139	1378
Hydro	944	907	840	908	932	795
Nuclear	..	..	..	..	..	..
Other	..	..	..	..	..	..
Own use in electricity, CHP and heat plants	70	70	67	75	79	83
Net production	1765	1862	1966	2076	2185	2284
Imports	..	..	..	..	184	388
Exports	..	..	..	..	..	..
Losses	356	377	404	421	486	555
Consumption	1409	1485	1562	1655	1883	2117
Energy industries own use	31	32	34	36	41	46
By industry and construction	372	394	414	439	500	561
By transport	6	6	6	7	7	8
By households and other cons.	1000	1053	1108	1173	1335	1502
Net installed capacity	Thousand kilowatts					
Total main activity and autoproducer	443	443	443	464	514	509
Combustible fuels	273	273	273	293	343	338
Hydro	170	170	170	170	170	170
Nuclear	..	..	..	..	..	..
Other	..	..	..	*1	*1	*1
Main activity	373	373	373	393	443	438
Combustible fuels	203	203	203	223	273	268
Hydro	170	170	170	170	170	170
Nuclear	..	..	..	..	..	..
Other	..	..	..	..	..	..
Combustible fuel input	Terajoules					
Gas-diesel oil	2236	2150	2838	3440	5074	6493
Fuel oil	646	162	40	40	40	81
Natural gas	8711	11801	13370	13397	10733	11865
Fuelwood	*271	*278	*285	*292	*299	*301
Total input	11864	14391	16533	17169	16146	18740
Total production	3208	3690	4295	4464	4788	5652
Estimated efficiency (% of production to input)	27	26	26	26	30	30

Statistics on electricity

Gambia

Item	2009	2010	2011	2012	2013	2014
Production, trade and consumption	**Million kilowatt-hours**					
Total main activity and autoproducer	242	*245	*256	*266	*261	*261
Combustible fuels	242	*245	*256	*266	*261	*261
Hydro	..	..	..	..	..	..
Nuclear	..	..	..	..	..	..
Other	..	..	..	..	..	..
Main activity	214	*217	*226	*236	*230	*230
Combustible fuels	214	*217	*226	*236	*230	*230
Hydro	..	..	..	..	..	..
Nuclear	..	..	..	..	..	..
Other	..	..	..	..	..	..
Own use in electricity, CHP and heat plants	6	*6	*4	*21	*12	-10
Net production	236	*239	*251	*245	*249	*270
Imports	..	..	..	..	..	..
Exports	..	..	..	..	..	..
Losses	*49	*53	*50	*52	*52	*52
Consumption	*187	*187	*199	*208	*209	*209
Energy industries own use	..	..	..	..	..	..
By industry and construction	32	*32	*34	*35	*35	*35
By transport	..	..	..	..	..	..
By households and other cons.	*155	*154	*166	*173	*174	*173
Net installed capacity	**Thousand kilowatts**					
Total main activity and autoproducer	62	*74	*74	*74	*91	116
Combustible fuels	62	*74	*74	*74	*91	114
Hydro	..	..	..	..	..	..
Nuclear	..	..	..	..	..	..
Other	..	..	..	..	..	2
Main activity	28	*40	*40	*40	*57	82
Combustible fuels	28	*40	*40	*40	*57	80
Hydro	..	..	..	..	..	..
Nuclear	..	..	..	..	..	..
Other	..	..	..	..	..	2
Combustible fuel input	**Terajoules**					
Gas-diesel oil	1634	*1892	*1978	*2021	*2021	*2150
Total input	1634	*1892	*1978	*2021	*2021	*2150
Total production	871	*883	*921	*959	*939	*939
Estimated efficiency (% of production to input)	53	47	47	47	46	44

Statistics on electricity

Georgia

Item	2009	2010	2011	2012	2013	2014
Production, trade and consumption	Million kilowatt-hours					
Total main activity and autoproducer	8408	10058	10104	9698	10059	10370
Combustible fuels	991	683	2212	2477	1788	2036
Hydro	7417	9375	7892	7221	8271	8334
Nuclear	..	..	..	..	..	..
Other	..	..	..	..	..	..
Main activity	8408	10058	10104	9698	10059	10370
Combustible fuels	991	683	2212	2477	1788	2036
Hydro	7417	9375	7892	7221	8271	8334
Nuclear	..	..	..	..	..	..
Other	..	..	..	..	..	..
Own use in electricity, CHP and heat plants	363	78	87	226	199	217
Net production	8045	9980	10017	9472	9860	10153
Imports	255	222	471	615	484	794
Exports	749	1524	931	528	450	545
Losses	1081	1103	1142	1094	805	600
Consumption	6563	7664	8511	8452	9089	9801
Energy industries own use	705	376	423	284	15	17
By industry and construction	1312	2137	2963	2876	2327	2816
By transport	348	546	373	405	282	267
By households and other cons.	4198	4605	4752	4887	6465	6701
Net installed capacity	Thousand kilowatts					
Total main activity and autoproducer	*4538	*4538	*4350	*4350	*4350	*4350
Combustible fuels	*1688	*1688	*1688	*1688	*1688	*1688
Hydro	*2850	*2850	*2662	*2662	*2662	*2662
Nuclear	..	..	..	..	..	..
Other	..	..	..	..	..	..
Main activity	*4538	*4538	*4350	*4350	*4350	*4350
Combustible fuels	*1688	*1688	*1688	*1688	*1688	*1688
Hydro	*2850	*2850	*2662	*2662	*2662	*2662
Nuclear	..	..	..	..	..	..
Other	..	..	..	..	..	..
Combustible fuel input	Terajoules					
Gas-diesel oil	2838	2322	516	0	0	0
Natural gas	19092	12869	19858	22588	16982	22345
Total input	21930	15191	20374	22588	16982	22345
Total production	3568	2459	7963	8917	6437	7330
Estimated efficiency (% of production to input)	16	16	39	39	38	33

Statistics on electricity

Germany

Item	2009	2010	2011	2012	2013	2014
Production, trade and consumption	**Million kilowatt-hours**					
Total main activity and autoproducer	595617	632983	613068	629812	638729	627795
Combustible fuels	388640	413207	410773	423424	428094	409688
Hydro	24682	27353	23511	27849	28782	25444
Nuclear	134932	140556	107971	99460	97290	97129
Other	47363	51867	70813	79079	84563	95534
Main activity	549657	580009	562454	585718	593837	582345
Combustible fuels	344545	362335	362124	381003	384628	365814
Hydro	24317	26975	23184	27499	28600	25282
Nuclear	134932	140556	107971	99460	97290	97129
Other	45863	50143	69175	77756	83319	94120
Own use in electricity, CHP and heat plants	37068	38151	36152	37072	36912	35843
Net production	558549	594832	576916	592740	601817	591952
Imports	41859	42962	51003	46268	39222	40435
Exports	54132	57917	54768	66810	71415	74320
Losses	25003	23974	24799	24562	24474	24159
Consumption	521272	555903	548352	547637	545150	533907
Energy industries own use	24013	23479	22806	21803	21949	21072
By industry and construction	202021	224530	229805	226239	224269	228773
By transport	11633	12119	12151	12084	11985	11594
By households and other cons.	283605	295775	283590	287511	286947	272468
Net installed capacity	**Thousand kilowatts**					
Total main activity and autoproducer	151678	162698	175896	177291	186117	198416
Combustible fuels	83245	85823	89421	89649	91368	97203
Hydro	11238	11218	11436	11257	11240	11234
Nuclear	20480	20467	20467	12068	12068	12074
Other	36715	45190	54572	64317	71441	77905
Main activity	141808	152795	166141	170272	176498	188609
Combustible fuels	73791	76351	80050	82912	82072	87747
Hydro	11158	11144	11367	11185	11198	11190
Nuclear	20480	20467	20467	12068	12068	12074
Other	36379	44833	54257	64107	71160	77598
Combustible fuel input	**Terajoules**					
Hard Coal	1069654	1149903	1075457	1130473	1254261	1162815
Brown Coal	1424132	1403420	1447201	1531986	1509182	1468554
Lignite briquettes	16692	16322	18453	20318	20807	18068
Biogas	104518	122294	148398	190358	214517	228266
Other oil products	43054	35255	25406	29386	28462	22110
Natural gas	780425	822592	771794	692263	616961	555763
Blast furnace gas	42027	72766	68853	67615	73193	69845
Fuelwood	*102379	*112509	*125705	*135851	*131071	*135878
Municipal waste	141740	152444	148154	156878	177360	189580
Industrial waste	20596	25697	25193	25478	19234	22477
Others	79857	71184	56029	64350	60028	56255
Total input	3825074	3984387	3910643	4044957	4105075	3929610
Total production	1399104	1487545	1478783	1524326	1541138	1474877
Estimated efficiency (% of production to input)	37	37	38	38	38	38

Statistics on electricity

Ghana

Item	2009	2010	2011	2012	2013	2014
Production, trade and consumption	Million kilowatt-hours					
Total main activity and autoproducer	8964	10167	11366	12204	13047	13139
Combustible fuels	2087	3171	3805	4133	4814	4752
Hydro	6877	6996	7561	8071	8233	8387
Nuclear	..	..	..	..	..	..
Other	..	..	..	..	..	..
Main activity	8958	10029	11200	12024	12867	12959
Combustible fuels	2081	3033	3639	3953	4634	4572
Hydro	6877	6996	7561	8071	8233	8387
Nuclear	..	..	..	..	..	..
Other	..	..	..	..	..	..
Own use in electricity, CHP and heat plants	6	0	166	180	177	176
Net production	8958	10167	11200	12024	12870	12963
Imports	198	106	81	128	27	51
Exports	752	1036	691	667	530	522
Losses	2088	1923	2113	2030	2183	2348
Consumption	6224	6877	7924	8890	9472	9522
Energy industries own use	..	..	..	..	..	..
By industry and construction	2920	3174	3901	4153	4435	4681
By transport	..	..	..	..	..	..
By households and other cons.	3304	3703	4023	4737	5037	4841
Net installed capacity	Thousand kilowatts					
Total main activity and autoproducer	2430	2600	2800	2910	3041	3041
Combustible fuels	1250	1420	1620	1730	1458	1458
Hydro	*1180	1180	1180	1180	1580	1580
Nuclear	..	..	..	..	..	..
Other	..	0	0	0	3	3
Main activity	*2200	2390	2590	2700	2831	2831
Combustible fuels	1020	1210	1410	1520	1248	1248
Hydro	*1180	1180	1180	1180	1580	1580
Nuclear	..	..	..	..	..	..
Other	..	0	0	0	3	3
Combustible fuel input	Terajoules					
Crude oil	22901	29631	11929	31941	36539	26357
Gas-diesel oil	..	..	284	606	219	65
Natural gas	..	10377	23404	12854	12140	24934
Total input	22901	40008	35617	45401	48898	51355
Total production	7513	11416	13698	14879	17330	17107
Estimated efficiency (% of production to input)	33	29	38	33	35	33

Statistics on electricity

Gibraltar

Item	2009	2010	2011	2012	2013	2014
Production, trade and consumption	**Million kilowatt-hours**					
Total main activity and autoproducer	174	177	171	176	189	198
Combustible fuels	174	177	171	176	189	198
Hydro	..	..	..	..	..	..
Nuclear	..	..	..	..	..	..
Other	..	..	..	..	..	..
Main activity	174	177	171	176	189	198
Combustible fuels	174	177	171	176	189	198
Hydro	..	..	..	..	..	..
Nuclear	..	..	..	..	..	..
Other	..	..	..	..	..	..
Own use in electricity, CHP and heat plants	3	4	3	4	4	4
Net production	171	173	168	172	185	194
Imports	..	..	..	..	..	..
Exports	..	..	..	..	..	..
Losses	5	5	5	5	6	6
Consumption	171	168	163	167	179	188
Energy industries own use	..	..	..	..	..	..
By industry and construction	..	..	..	..	..	..
By transport	..	..	..	..	..	..
By households and other cons.	171	168	163	167	179	188
Net installed capacity	**Thousand kilowatts**					
Total main activity and autoproducer	43	43	43	43	43	43
Combustible fuels	43	43	43	43	43	43
Hydro	..	..	..	..	..	..
Nuclear	..	..	..	..	..	..
Other	..	..	..	..	..	..
Main activity	43	43	43	43	43	43
Combustible fuels	43	43	43	43	43	43
Hydro	..	..	..	..	..	..
Nuclear	..	..	..	..	..	..
Other	..	..	..	..	..	..
Combustible fuel input	**Terajoules**					
Fuel oil	1737	1737	1697	1737	1858	1939
Total input	1737	1737	1697	1737	1858	1939
Total production	626	637	616	634	680	713
Estimated efficiency (% of production to input)	36	37	36	36	37	37

Statistics on electricity

Greece

Item	2009	2010	2011	2012	2013	2014
Production, trade and consumption	**Million kilowatt-hours**					
Total main activity and autoproducer	61365	57392	59436	60959	57152	50474
Combustible fuels	53127	47035	51236	50824	42981	38386
Hydro	5645	7485	4275	4591	6384	4607
Nuclear	..	..	..	..	..	..
Other	2593	2872	3925	5544	7787	7481
Main activity	59407	54900	56883	58608	55267	48682
Combustible fuels	51169	44543	48683	48473	41096	36594
Hydro	5645	7485	4275	4591	6384	4607
Nuclear	..	..	..	..	..	..
Other	2593	2872	3925	5544	7787	7481
Own use in electricity, CHP and heat plants	5292	4003	5523	7305	4592	3772
Net production	56073	53389	53913	53654	52560	46702
Imports	7600	8517	7180	5954	5788	9461
Exports	3233	2811	3948	4169	3901	642
Losses	3223	3783	2820	1611	3895	4149
Consumption	57217	55312	54325	53828	50552	51372
Energy industries own use	2504	2192	2531	1811	1761	1872
By industry and construction	14067	14142	14640	11573	11366	12869
By transport	232	183	184	189	267	343
By households and other cons.	40414	38795	36970	40255	37158	36288
Net installed capacity	**Thousand kilowatts**					
Total main activity and autoproducer	14627	15312	16524	17751	18855	18895
Combustible fuels	10209	10597	11048	11226	11229	10932
Hydro	3201	3215	3224	3236	3238	3389
Nuclear	..	..	..	..	..	..
Other	1217	1500	2252	3289	4388	4574
Main activity	14085	14790	15984	17229	18328	18390
Combustible fuels	9667	10075	10508	10704	10702	10427
Hydro	3201	3215	3224	3236	3238	3389
Nuclear	..	..	..	..	..	..
Other	1217	1500	2252	3289	4388	4574
Combustible fuel input	**Terajoules**					
Hard Coal	520	4250	813	54	27	..
Brown Coal	343680	312380	320533	331017	283513	271373
Biogas	2288	1981	2366	3072	2975	2955
Gas-diesel oil	14534	13932	12943	11438	10019	11782
Fuel oil	54459	38986	37128	37774	34623	35794
Liquefied petroleum gas	0	142	1750	1419	189	142
Refinery gas	9207	9603	9356	10692	9455	11435
Natural gas	84511	95857	111198	103109	91617	59593
Industrial waste	169	1341	1175	625	883	868
Total input	509368	478472	497261	499200	433301	393941
Total production	191257	169326	184450	182966	154732	138190
Estimated efficiency (% of production to input)	38	35	37	37	36	35

Statistics on electricity

Greenland

Item	2009	2010	2011	2012	2013	2014
Production, trade and consumption	**Million kilowatt-hours**					
Total main activity and autoproducer	573	533	565	574	568	580
Combustible fuels	308	237	227	229	191	184
Hydro	265	296	338	345	378	396
Nuclear	..	..	..	..	..	..
Other	..	..	..	..	..	..
Main activity	419	415	452	459	473	488
Combustible fuels	154	119	114	115	95	92
Hydro	265	296	338	345	378	396
Nuclear	..	..	..	..	..	..
Other	..	..	..	..	..	..
Own use in electricity, CHP and heat plants	241	189	207	257	226	399
Net production	333	344	358	317	342	181
Imports	..	..	..	..	..	..
Exports	..	..	..	..	..	..
Losses	11	15	22	7	22	12
Consumption	306	300	320	314	318	331
Energy industries own use	..	..	..	..	..	..
By industry and construction	64	64	62	58	60	60
By transport	..	..	..	..	..	..
By households and other cons.	242	236	258	257	259	270
Net installed capacity	**Thousand kilowatts**					
Total main activity and autoproducer	*174	*174	*174	*172	*187	187
Combustible fuels	*106	*106	*106	*96	*96	96
Hydro	68	68	*68	76	91	91
Nuclear	..	..	..	..	..	..
Other	..	..	..	..	..	..
Main activity	*158	*158	*158	*156	171	171
Combustible fuels	*90	*90	*90	*80	*80	80
Hydro	68	68	*68	76	91	91
Nuclear	..	..	..	..	..	..
Other	..	..	..	..	..	..
Combustible fuel input	**Terajoules**					
Gas-diesel oil	1615	1365	1332	1384	1146	1179
Municipal waste	80	90	91	103	117	92
Total input	1695	1455	1422	1487	1264	1271
Total production	1108	854	818	826	686	661
Estimated efficiency (% of production to input)	65	59	58	56	54	52

Statistics on electricity

Grenada

Item	2009	2010	2011	2012	2013	2014
Production, trade and consumption	Million kilowatt-hours					
Total main activity and autoproducer	203	209	204	200	197	200
Combustible fuels	203	209	204	200	197	200
Hydro	..	..	..	..	..	..
Nuclear	..	..	..	..	..	..
Other	..	..	..	..	..	..
Main activity	203	209	204	200	197	200
Combustible fuels	203	209	204	200	197	200
Hydro	..	..	..	..	..	..
Nuclear	..	..	..	..	..	..
Other	..	..	..	..	..	..
Own use in electricity, CHP and heat plants	8	7	7	7	6	6
Net production	195	201	197	193	191	194
Imports	..	..	..	..	..	..
Exports	..	..	..	..	..	..
Losses	18	17	16	15	15	15
Consumption	177	185	181	179	176	179
Energy industries own use	..	..	..	..	..	..
By industry and construction	6	6	6	6	6	6
By transport	..	..	..	..	..	..
By households and other cons.	172	179	175	173	170	173
Net installed capacity	Thousand kilowatts					
Total main activity and autoproducer	49	49	50	49	49	49
Combustible fuels	49	49	50	49	49	49
Hydro	..	..	..	..	..	..
Nuclear	..	..	..	..	..	..
Other	..	..	..	..	..	..
Main activity	49	49	50	49	49	49
Combustible fuels	49	49	50	49	49	49
Hydro	..	..	..	..	..	..
Nuclear	..	..	..	..	..	..
Other	..	..	..	..	..	..
Combustible fuel input	Terajoules					
Gas-diesel oil	1757	1861	1811	2099	2558	*1725
Total input	1757	1861	1811	2099	2558	*1725
Total production	731	751	734	719	708	720
Estimated efficiency (% of production to input)	42	40	41	34	28	42

85

Statistics on electricity

Guadeloupe

Item	2009	2010	2011	2012	2013	2014
Production, trade and consumption	**Million kilowatt-hours**					
Total main activity and autoproducer	*1709	*1809	*1767	1817	1818	1809
Combustible fuels	*1554	*1634	*1555	1610	1562	1563
Hydro	*14	*14	*14	14	19	22
Nuclear	..	..	..	..	..	..
Other	*141	*161	*198	193	237	224
Main activity	*1709	*1809	*1767	1817	1818	1809
Combustible fuels	*1554	*1634	*1555	1610	1562	1563
Hydro	*14	*14	*14	14	19	22
Nuclear	..	..	..	..	..	..
Other	*141	*161	*198	193	237	224
Own use in electricity, CHP and heat plants	*81	*79	*75	91	89	75
Net production	*1628	*1730	*1692	1726	1729	1734
Imports	..	..	..	..	..	..
Exports	..	..	..	..	..	..
Losses	*200	*200	*200	219	219	207
Consumption	*1428	*1530	*1492	1507	1510	1508
Energy industries own use	..	..	..	..	..	..
By industry and construction	*70	*70	*70	64	66	86
By transport	..	..	..	..	..	..
By households and other cons.	*1358	*1460	*1422	1443	1444	1422
Net installed capacity	**Thousand kilowatts**					
Total main activity and autoproducer	441	440	467	493	496	496
Combustible fuels	368	359	359	380	380	380
Hydro	10	8	8	9	9	9
Nuclear	..	..	..	..	..	..
Other	63	*73	*100	105	108	108
Main activity	441	440	467	493	496	496
Combustible fuels	368	359	359	380	380	380
Hydro	10	8	8	9	9	9
Nuclear	..	..	..	..	..	..
Other	63	*73	*100	105	108	108
Combustible fuel input	**Terajoules**					
Hard Coal	*4257	*3999	*5160	*5160	*5160	*5160
Gas-diesel oil	*4730	*4902	*4945	*6235	*6020	*5805
Fuel oil	*7312	*7393	*7393	*7393	*7434	*7474
Bagasse	*1081	*1158	*1197	*1236	*1236	*1236
Total input	*17380	*17453	*18695	*20024	*19849	*19675
Total production	*5594	*5882	*5598	5796	5623	5627
Estimated efficiency (% of production to input)	32	34	30	29	28	29

Statistics on electricity

Guam

Item	2009	2010	2011	2012	2013	2014
Production, trade and consumption	**Million kilowatt-hours**					
Total main activity and autoproducer	1868	*1882	*1858	*1799	1736	*1750
Combustible fuels	1868	*1882	*1858	*1799	1736	*1750
Hydro	..	..	..	..	..	..
Nuclear	..	..	..	..	..	..
Other	..	..	..	..	..	..
Main activity	1868	*1882	*1858	*1799	1736	*1750
Combustible fuels	1868	*1882	*1858	*1799	1736	*1750
Hydro	..	..	..	..	..	..
Nuclear	..	..	..	..	..	..
Other	..	..	..	..	..	..
Own use in electricity, CHP and heat plants	105	*113	*112	*109	86	*87
Net production	1763	*1769	*1746	*1690	1650	*1663
Imports	..	..	..	..	..	..
Exports	..	..	..	..	..	..
Losses	*138	*132	*128	*127	101	*85
Consumption	1625	1638	1618	1564	1550	*1578
Energy industries own use	..	..	..	..	..	..
By industry and construction	..	..	..	..	..	..
By transport	..	..	..	..	..	..
By households and other cons.	1625	1638	1618	1564	1550	*1578
Net installed capacity	**Thousand kilowatts**					
Total main activity and autoproducer	552	552	552	552	552	*552
Combustible fuels	552	552	552	552	552	*552
Hydro	..	..	..	..	..	..
Nuclear	..	..	..	..	..	..
Other	..	..	..	..	..	..
Main activity	552	552	552	552	552	*552
Combustible fuels	552	552	552	552	552	*552
Hydro	..	..	..	..	..	..
Nuclear	..	..	..	..	..	..
Other	..	..	..	..	..	..
Combustible fuel input	**Terajoules**					
Total input	..	..	..	..	..	..
Total production	6724	*6777	*6689	*6476	6251	*6300
Estimated efficiency (% of production to input)	..	..	..	..	..	..

Statistics on electricity

Guatemala

Item	2009	2010	2011	2012	2013	2014
Production, trade and consumption	**Million kilowatt-hours**					
Total main activity and autoproducer	9047	8893	8798	9335	10024	10638
Combustible fuels	5718	4773	4376	4631	5152	5531
Hydro	2942	3849	4175	4459	4660	4853
Nuclear	..	..	..	..	..	..
Other	387	271	246	246	212	254
Main activity	8323	8234	8189	8702	9170	9782
Combustible fuels	5016	4136	3790	4020	4326	4703
Hydro	2920	3827	4153	4436	4632	4825
Nuclear						
Other	387	271	246	246	212	254
Own use in electricity, CHP and heat plants	228	381	344	393	385	421
Net production	8819	8512	8454	8942	9639	10217
Imports	37	362	511	372	313	545
Exports	94	139	193	346	669	1053
Losses	1291	1187	1259	1226	1201	1015
Consumption	7464	7549	7688	7741	7582	8288
Energy industries own use	..	..	..	..	..	..
By industry and construction	3014	3054	3122	3126	3062	3347
By transport	..	..	..	..	..	..
By households and other cons.	4450	4495	4566	4615	4520	4941
Net installed capacity	**Thousand kilowatts**					
Total main activity and autoproducer	2382	2454	2510	2796	2974	2610
Combustible fuels	1555	1552	1570	1773	1923	1580
Hydro	778	853	891	991	1002	991
Nuclear	..	..	..	..	..	..
Other	49	49	49	32	49	39
Main activity	2007	2072	2126	2362	2375	2137
Combustible fuels	1184	1174	1190	1344	1328	1111
Hydro	774	849	887	987	998	987
Nuclear	..	..	..	..	..	..
Other	49	49	49	32	49	39
Combustible fuel input	**Terajoules**					
Hard Coal	7585	12668	12694	12823	12828	18581
Gas-diesel oil	2193	43	344	86	340	340
Fuel oil	30058	17412	16524	15716	19331	19610
Bagasse	*50288	*46597	*72189	*75799	80657	80657
Total input	*90124	*76720	*101750	*104423	113156	119188
Total production	20585	17182	15754	16671	18548	19911
Estimated efficiency (% of production to input)	23	22	15	16	16	17

Statistics on electricity

Guernsey

Item	2009	2010	2011	2012	2013	2014
Production, trade and consumption	**Million kilowatt-hours**					
Total main activity and autoproducer	174	152	85	68	289	227
Combustible fuels	174	152	85	68	289	227
Hydro	..	..	..	..	..	..
Nuclear	..	..	..	..	..	..
Other	..	..	..	..	..	..
Main activity	174	152	85	68	289	227
Combustible fuels	174	152	85	68	289	227
Hydro	..	..	..	..	..	..
Nuclear	..	..	..	..	..	..
Other	..	..	..	..	..	..
Own use in electricity, CHP and heat plants	0	0	0	0	0	0
Net production	174	152	85	68	289	227
Imports	210	239	309	313	111	156
Exports	..	..	..	..	..	..
Losses	*29	*29	*18	*18	*22	*21
Consumption	355	362	375	*363	*378	*362
Energy industries own use	..	..	..	..	..	..
By industry and construction	43	48	53	*50	*54	*51
By transport	..	..	..	..	..	..
By households and other cons.	312	315	322	*313	*325	*311
Net installed capacity	**Thousand kilowatts**					
Total main activity and autoproducer	115	115	115	*115	*115	*115
Combustible fuels	115	115	115	*115	*115	*115
Hydro	..	..	..	..	..	..
Nuclear	..	..	..	..	..	..
Other	..	..	..	..	..	..
Main activity	115	115	115	*115	*115	*115
Combustible fuels	115	115	115	*115	*115	*115
Hydro	..	..	..	..	..	..
Nuclear	..	..	..	..	..	..
Other	..	..	..	..	..	..
Combustible fuel input	**Terajoules**					
Gas-diesel oil	14	13	14	*49	*49	*49
Fuel oil	1443	1255	692	*606	*2424	*1903
Total input	1457	1267	706	*655	*2473	*1951
Total production	625	548	305	245	1040	818
Estimated efficiency (% of production to input)	43	43	43	37	42	42

Guinea

Item	2009	2010	2011	2012	2013	2014
Production, trade and consumption	Million kilowatt-hours					
Total main activity and autoproducer	*1004	*958	*890	*1071	*995	*1015
Combustible fuels	*542	*484	*476	*580	*513	*525
Hydro	462	474	414	491	482	*490
Nuclear	..	..	..	..	..	..
Other	..	..	..	..	..	..
Main activity	661	615	548	729	653	*665
Combustible fuels	199	141	134	238	171	*175
Hydro	462	474	414	491	482	*490
Nuclear	..	..	..	..	..	..
Other	..	..	..	..	..	..
Own use in electricity, CHP and heat plants	*100	*96	*89	*108	*99	*101
Net production	*904	*862	*801	*963	*896	*914
Imports	..	..	..	..	..	..
Exports	..	..	..	..	..	..
Losses	*180	*167	*154	*197	*170	*173
Consumption	*724	*694	*647	*766	*726	*741
Energy industries own use	..	..	..	..	..	..
By industry and construction	*350	*352	*333	*345	*353	*355
By transport	..	..	..	..	..	..
By households and other cons.	*374	*342	*314	*421	*373	*386
Net installed capacity	Thousand kilowatts					
Total main activity and autoproducer	*423	*432	*385	*394	518	*518
Combustible fuels	*274	*283	*236	*246	370	*370
Hydro	149	149	149	149	149	*149
Nuclear	..	..	..	..	..	..
Other	..	..	..	..	..	..
Main activity	243	252	205	214	338	*338
Combustible fuels	115	124	77	87	211	*211
Hydro	128	128	128	128	128	*128
Nuclear	..	..	..	..	..	..
Other	..	..	..	..	..	..
Combustible fuel input	Terajoules					
Gas-diesel oil	0	0	0	*3010	*2150	*2150
Fuel oil	*6747	*5898	*5777	*4040	*3636	*4242
Total input	*6747	*5898	*5777	*7050	*5786	*6392
Total production	*1952	*1743	*1714	*2089	*1848	*1890
Estimated efficiency (% of production to input)	29	30	30	30	32	30

Statistics on electricity

Guinea-Bissau

Item	2009	2010	2011	2012	2013	2014
Production, trade and consumption	**Million kilowatt-hours**					
Total main activity and autoproducer	*31	*32	*33	*34	*35	*35
Combustible fuels	*31	*32	*33	*34	*35	*35
Hydro	..	..	..	..	..	..
Nuclear	..	..	..	..	..	..
Other	..	..	..	..	..	..
Main activity	*24	*25	*26	*26	*27	*27
Combustible fuels	*24	*25	*26	*26	*27	*27
Hydro	..	..	..	..	..	..
Nuclear	..	..	..	..	..	..
Other	..	..	..	..	..	..
Own use in electricity, CHP and heat plants	0	0	0	0	0	0
Net production	*31	*32	*33	*34	*35	*35
Imports	..	..	..	..	..	..
Exports	..	..	..	..	..	..
Losses	..	..	..	..	..	..
Consumption	*31	*32	*33	*34	*35	*35
Energy industries own use	..	..	..	..	..	..
By industry and construction	*18	*18	*19	*20	*20	*20
By transport	..	..	..	..	..	..
By households and other cons.	*13	*14	*14	*15	*15	*15
Net installed capacity	**Thousand kilowatts**					
Total main activity and autoproducer	27	*27	*27	*27	*27	*28
Combustible fuels	27	*27	*27	*27	*27	*28
Hydro	..	..	..	..	..	..
Nuclear	..	..	..	..	..	..
Other	..	..	..	..	..	..
Main activity	25	*25	*25	*25	*25	*26
Combustible fuels	25	*25	*25	*25	*25	*26
Hydro	..	..	..	..	..	..
Nuclear	..	..	..	..	..	..
Other	..	..	..	..	..	..
Combustible fuel input	**Terajoules**					
Gas-diesel oil	*323	*340	*344	357	*366	*387
Total input	*323	*340	*344	357	*366	*387
Total production	*111	*115	*119	*123	*127	*127
Estimated efficiency (% of production to input)	34	34	35	35	35	33

Statistics on electricity

Guyana

Item	2009	2010	2011	2012	2013	2014
Production, trade and consumption	**Million kilowatt-hours**					
Total main activity and autoproducer	723	868	910	942	962	979
Combustible fuels	723	868	910	942	962	979
Hydro	..	..	..	..	..	..
Nuclear	..	..	..	..	..	..
Other	..	..	..	..	..	..
Main activity	586	568	605	658	660	778
Combustible fuels	586	568	605	658	660	778
Hydro	..	..	..	..	..	..
Nuclear	..	..	..	..	..	..
Other	..	..	..	..	..	..
Own use in electricity, CHP and heat plants	25	126	130	128	28	29
Net production	699	742	780	815	934	951
Imports	..	..	..	..	..	..
Exports	..	..	..	..	..	..
Losses	216	252	263	242	230	215
Consumption	483	477	502	526	704	736
Energy industries own use	..	..	..	..	..	..
By industry and construction	51	147	153	158	235	306
By transport	..	..	..	..	..	..
By households and other cons.	431	330	348	368	469	430
Net installed capacity	**Thousand kilowatts**					
Total main activity and autoproducer	342	348	349	363	*383	*380
Combustible fuels	342	348	349	363	*383	*380
Hydro	..	..	..	..	..	..
Nuclear	..	..	..	..	..	..
Other	..	..	..	..	..	..
Main activity	163	170	170	182	*192	*190
Combustible fuels	163	170	170	182	*192	*190
Hydro	..	..	..	..	..	..
Nuclear	..	..	..	..	..	..
Other	..	..	..	..	..	..
Combustible fuel input	**Terajoules**					
Gas-diesel oil	2576	2907	3840	2911	3178	2884
Fuel oil	4723	7022	5979	6832	7329	7280
Bagasse	364	246	279	425	513	480
Total input	7662	10175	10098	10167	11019	10643
Total production	2603	3124	3277	3393	3464	3526
Estimated efficiency (% of production to input)	34	31	32	33	31	33

Statistics on electricity

Haiti

Item	2009	2010	2011	2012	2013	2014
Production, trade and consumption	**Million kilowatt-hours**					
Total main activity and autoproducer	721	587	974	1100	1055	1033
Combustible fuels	514	410	853	879	914	943
Hydro	207	177	121	221	141	90
Nuclear	..	..	..	..	..	..
Other	..	..	..	..	..	..
Main activity	721	587	943	1052	1005	943
Combustible fuels	514	410	822	831	864	853
Hydro	207	177	121	221	141	90
Nuclear	..	..	..	..	..	..
Other	..	..	..	..	..	..
Own use in electricity, CHP and heat plants	17	15	59	28	19	11
Net production	704	572	915	1072	1036	1022
Imports	..	..	..	..	..	..
Exports	..	..	..	..	..	..
Losses	369	343	627	627	599	621
Consumption	336	228	235	493	456	410
Energy industries own use	..	..	..	..	..	..
By industry and construction	110	75	83	194	171	184
By transport	..	..	..	..	..	..
By households and other cons.	226	153	152	299	285	226
Net installed capacity	**Thousand kilowatts**					
Total main activity and autoproducer	240	267	267	267	*340	*320
Combustible fuels	178	207	207	207	*262	*259
Hydro	62	61	61	61	77	61
Nuclear	..	..	..	..	..	..
Other	..	..	..	..	..	..
Main activity	217	193	193	193	*246	*227
Combustible fuels	155	133	133	133	*168	*166
Hydro	62	61	61	61	77	61
Nuclear	..	..	..	..	..	..
Other	..	..	..	..	..	..
Combustible fuel input	**Terajoules**					
Gas-diesel oil	*2709	2881	*4300	6880	6149	5504
Fuel oil	1010	808	*2020	2262	2666	5171
Total input	*3719	3689	*6320	9142	8815	10675
Total production	1850	1476	3071	3164	3290	3395
Estimated efficiency (% of production to input)	50	40	49	35	37	32

Statistics on electricity

Honduras

Item	2009	2010	2011	2012	2013	2014
Production, trade and consumption	**Million kilowatt-hours**					
Total main activity and autoproducer	6625	6777	7653	7740	8074	8038
Combustible fuels	3828	3697	4721	4615	5025	5038
Hydro	2797	3080	2815	2787	2739	2602
Nuclear	..	..			..	..
Other	..	..	117	338	310	398
Main activity	6613	6765	7170	7231	7543	7687
Combustible fuels	3816	3685	4238	4106	4494	4687
Hydro	2797	3080	2815	2787	2739	2602
Nuclear	..	..			..	..
Other	..	..	117	338	310	398
Own use in electricity, CHP and heat plants	12	12	14	238	248	0
Net production	6613	6765	7639	7502	7826	8038
Imports	1	22	171	154	268	827
Exports	46	13	127	79	156	505
Losses	1491	1647	2066	2314	2527	2809
Consumption	5037	5100	5449	5308	5455	5552
Energy industries own use	..	..			..	..
By industry and construction	1237	1267	1590	1726	1787	1548
By transport	..	..			..	..
By households and other cons.	3800	3833	3859	3582	3668	4004
Net installed capacity	**Thousand kilowatts**					
Total main activity and autoproducer	1606	1611	1722	1783	1806	1915
Combustible fuels	1084	1084	1089	1143	1146	1139
Hydro	522	527	531	538	558	624
Nuclear	..	..			..	..
Other	..	0	102	102	102	152
Main activity	1530	1535	1594	1624	1648	1734
Combustible fuels	1010	1010	961	984	988	958
Hydro	520	525	531	538	558	624
Nuclear	..	..			..	..
Other	..	0	102	102	102	152
Combustible fuel input	**Terajoules**					
Hard Coal	0	0	1006	929	929	413
Gas-diesel oil	989	1032	1892	1161	2279	6794
Fuel oil	28765	28320	34582	21493	25048	39471
Bagasse	*2419	2304	9331	9951	10454	7562
Total input	32173	31656	46812	33534	38710	54239
Total production	13781	13309	16996	16614	18090	18137
Estimated efficiency (% of production to input)	43	42	36	50	47	33

Statistics on electricity

Hungary

Item	2009	2010	2011	2012	2013	2014
Production, trade and consumption	Million kilowatt-hours					
Total main activity and autoproducer	35908	37371	36019	34635	30291	29371
Combustible fuels	19922	20887	19485	17851	13933	12652
Hydro	228	188	222	213	213	302
Nuclear	15426	15761	15685	15793	15370	15649
Other	332	535	627	778	775	768
Main activity	35555	36945	35323	34119	29679	28729
Combustible fuels	19570	20462	18790	17343	13378	12121
Hydro	228	188	222	213	213	302
Nuclear	15426	15761	15685	15793	15370	15649
Other	331	534	626	770	718	657
Own use in electricity, CHP and heat plants	2564	2758	2486	2284	2246	2240
Net production	33344	34613	33533	32351	28045	27131
Imports	10972	9897	14664	16970	16635	19079
Exports	5459	4702	8021	9003	4758	5689
Losses	3604	3801	3784	3684	3663	3631
Consumption	35253	36007	36358	34247	36280	35866
Energy industries own use	2103	1800	1818	1435	1407	1162
By industry and construction	8561	9784	9878	8910	14840	14678
By transport	1201	1106	1117	983	1227	1146
By households and other cons.	23388	23317	23545	22919	18806	18880
Net installed capacity	Thousand kilowatts					
Total main activity and autoproducer	8867	8993	9654	9400	8418	8809
Combustible fuels	6670	6645	7264	7007	5987	6333
Hydro	53	53	55	56	57	57
Nuclear	1940	2000	2000	2000	2000	2000
Other	204	295	335	337	374	419
Main activity	8743	8843	9483	9287	8217	8547
Combustible fuels	6547	6497	7097	6906	5831	6161
Hydro	53	53	55	56	57	57
Nuclear	1940	2000	2000	2000	2000	2000
Other	203	293	331	325	329	329
Combustible fuel input	Terajoules					
Brown Coal	67813	67372	71632	70166	68609	64078
Biogas	537	845	1931	1608	2269	2115
Gas-diesel oil	1591	1677	731	645	129	172
Fuel oil	5252	1495	646	929	566	525
Natural gas	111963	120759	106501	92333	57652	44564
Coke-oven gas	507	927	1895	2162	1596	1786
Blast furnace gas	2360	3341	3038	2975	1246	2141
Fuelwood	*26903	*27094	*19982	*16471	*17663	*21531
Municipal waste	3860	4458	4030	3842	3923	4096
Industrial waste	46	115	578	341	296	398
Others	6693	8096	879	813	90	119
Total input	227525	236178	211844	192285	154039	141526
Total production	71719	75193	70146	64264	50159	45547
Estimated efficiency (% of production to input)	32	32	33	33	33	32

Statistics on electricity

Iceland

Item	2009	2010	2011	2012	2013	2014
Production, trade and consumption	**Million kilowatt-hours**					
Total main activity and autoproducer	16834	17059	17211	17549	18116	18122
Combustible fuels	2	2	2	3	5	3
Hydro	12279	12592	12507	12337	12863	12873
Nuclear	..	..	..	..	..	..
Other	4553	4465	4702	5209	5248	5246
Main activity	16834	17059	17211	17549	18116	18122
Combustible fuels	2	2	2	3	5	3
Hydro	12279	12592	12507	12337	12863	12873
Nuclear	..	..	..	..	..	..
Other	4553	4465	4702	5209	5248	5246
Own use in electricity, CHP and heat plants	313	313	360	329	342	449
Net production	16521	16746	16851	17220	17774	17673
Imports	..	..	..	..	..	..
Exports	..	..	..	..	..	..
Losses	508	699	503	486	374	498
Consumption	15825	15861	16152	16523	16989	17351
Energy industries own use	166	154	142	166	166	521
By industry and construction	13502	13498	13868	14201	14678	14682
By transport	0	0	0	1	2	4
By households and other cons.	2157	2209	2142	2155	2143	2144
Net installed capacity	**Thousand kilowatts**					
Total main activity and autoproducer	2571	2579	2669	2656	2765	2766
Combustible fuels	121	121	120	114	114	114
Hydro	1875	1883	1884	1877	1984	1984
Nuclear	..	..	..	..	..	..
Other	575	575	665	665	667	668
Main activity	2571	2579	2669	2656	2765	2766
Combustible fuels	121	121	120	114	114	114
Hydro	1875	1883	1884	1877	1984	1984
Nuclear	..	..	..	..	..	..
Other	575	575	665	665	667	668
Combustible fuel input	**Terajoules**					
Biogas	2	0	0	0	0	0
Gas-diesel oil	43	43	43	43	43	43
Municipal waste	8	0	0	0	0	0
Total input	53	43	43	43	43	43
Total production	7	7	7	11	18	11
Estimated efficiency (% of production to input)	14	17	17	25	42	25

96

Statistics on electricity

India

Item	2009	2010	2011	2012	2013	2014
Production, trade and consumption	**Million kilowatt-hours**					
Total main activity and autoproducer	899389	954539	973006	1035264	1193480	1308873
Combustible fuels	755656	795052	813625	860173	980599	1111341
Hydro	106909	114486	103741	113855	141637	129353
Nuclear	18636	26266	32287	32871	34228	36102
Other	18188	18735	23353	28365	37016	32077
Main activity	796281	844846	844605	906863	1002743	1146816
Combustible fuels	652777	685588	685453	732001	792477	951504
Hydro	106680	114257	103512	113626	141508	129244
Nuclear	18636	26266	32287	32871	34228	36102
Other	18188	18735	23353	28365	34530	29966
Own use in electricity, CHP and heat plants	49706	52380	57238	*57238	81588	29966
Net production	849683	902159	915768	978026	1111892	1278907
Imports	5359	5610	8441	5152	5598	4998
Exports	519	584	0	0	0	0
Losses	180321	146063	185193	185306	220257	*241550
Consumption	682144	694392	772603	785130	897233	955927
Energy industries own use	..	..	..	..	7178	*7600
By industry and construction	326252	272589	346469	352291	374278	*418346
By transport	12408	14003	14327	14206	15447	16177
By households and other cons.	343484	407800	411807	418633	500330	513804
Net installed capacity	**Thousand kilowatts**					
Total main activity and autoproducer	198798	219591	252471	278275	*294873	*316379
Combustible fuels	146388	164118	192556	215522	*226449	*241396
Hydro	36924	37628	39051	39552	*40061	*40061
Nuclear	4560	4780	4780	4780	4780	5780
Other	10926	13065	16084	18421	*23583	29142
Main activity	170324	186691	215961	241765	*256786	*271722
Combustible fuels	117975	131279	156107	179073	*190000	*198745
Hydro	36863	37567	38990	39491	*40000	*40000
Nuclear	4560	4780	4780	4780	4780	5780
Other	10926	13065	16084	18421	*22006	27197
Combustible fuel input	**Terajoules**					
Hard Coal	7698132	6990453	8176582	8841855	8249905	8625865
Brown Coal	270029	285416	306073	356142	*346863	376809
Gas-diesel oil	13029	7138	7224	9202	290293	*281048
Fuel oil	184709	33249	26139	*19554	34906	84234
Liquefied petroleum gas	..	..	..	..	142	142
Naphtha	28436	18646	8322	15219	9568	8856
Natural gas	833684	920232	737965	501381	642401	418305
Coke-oven gas	19215	20340	20317	23260	25477	*21800
Fuelwood	*127644	*157824	*189684	*221496	*221900	*222700
Vegetal waste	*127644	*157824	*189684	*221496	*225681	*230799
Total input	9302520	8591122	9661990	10209605	10047135	10270557
Total production	2720362	2862187	2929050	3096623	3530156	4000828
Estimated efficiency (% of production to input)	29	33	30	30	35	39

Indonesia

Item	2009	2010	2011	2012	2013	2014
Production, trade and consumption	**Million kilowatt-hours**					
Total main activity and autoproducer	157516	169570	182389	196181	216186	220794
Combustible fuels	136833	142752	160593	173957	189835	191991
Hydro	11384	17456	12419	12799	16930	18800
Nuclear	..	..				..
Other	9299	9362	9377	9425	9421	10003
Main activity	155721	151129	148622	155615	169036	175297
Combustible fuels	136119	125944	128934	135670	146602	150478
Hydro	10307	15827	10316	10525	13014	14816
Nuclear	..	..	..	..	..	..
Other	9295	9358	9372	9420	9420	10003
Own use in electricity, CHP and heat plants	5224	6273	6887	5791	5011	5004
Net production	152292	163297	175502	190390	211175	215790
Imports	1262	2224	2542	2440	2990	8990
Exports	..	..	..	..	..	..
Losses	15210	15953	16672	17847	20701	21423
Consumption	137597	152805	160517	173991	195198	170088
Energy industries own use	..	..	..		..	
By industry and construction	49219	56083	55375	60176	72153	36737
By transport	111	89	88	108	129	155
By households and other cons.	88267	96633	105054	113707	122916	133196
Net installed capacity	**Thousand kilowatts**					
Total main activity and autoproducer	36775	38506	47756	53125	55426	60588
Combustible fuels	30792	32315	42163	46603	48875	53692
Hydro	4876	5054	4406	5067	5054	5306
Nuclear	..	..	..	..	..	..
Other	*1107	1138	1186	1454	1497	1590
Main activity	30863	32792	40832	45023	46489	51651
Combustible fuels	26251	27972	35743	39020	40441	45258
Hydro	3508	3686	3906	4552	4553	4805
Nuclear	..	..	..	..	..	..
Other	*1104	1135	1182	1450	1495	1588
Combustible fuel input	**Terajoules**					
Hard Coal	563972	715146	1156713	1364290	1562328	1711643
Brown Coal	*301046	*161821	*160120	*176000	*176000	*167368
Gas-diesel oil	221708	270556	335615	322113	258129	253528
Fuel oil	89526	62418	95869	60600	44521	24725
Natural gas	288561	294735	383947	479651	533975	509724
Fuelwood	*1134	*1674	*3006	*2592	*4104	*3708
Charcoal	..	..	..	..	2183	..
Total input	1465947	1506351	2135270	2405246	2581240	2670696
Total production	492599	513907	578135	626245	683406	691168
Estimated efficiency (% of production to input)	34	34	27	26	26	26

Iran (Islamic Rep. of)

Item	2009	2010	2011	2012	2013	2014
Production, trade and consumption	Million kilowatt-hours					
Total main activity and autoproducer	221372	232965	240052	254276	262434	274609
Combustible fuels	213912	223266	227428	239752	242909	255917
Hydro	7233	9526	12058	12447	14582	13862
Nuclear	..	..	327	1847	4546	4472
Other	227	173	239	230	397	358
Main activity	213822	225386	230216	243536	255885	268339
Combustible fuels	206362	215687	217592	229012	236360	249647
Hydro	7233	9526	12058	12447	14582	13862
Nuclear	..	..	327	1847	4546	4472
Other	227	173	239	230	397	358
Own use in electricity, CHP and heat plants	8487	8100	8482	8558	9137	8889
Net production	212885	224865	231570	245718	253297	265720
Imports	2068	3015	3656	3897	3707	3772
Exports	6152	6707	8668	11029	11586	9660
Losses	34547	33062	34906	36794	38049	34610
Consumption	174253	188141	191698	201793	208335	223957
Energy industries own use	*2051	2083	2240	2391	2705	2546
By industry and construction	58368	63060	69143	71989	72852	75689
By transport	282	300	354	371	323	363
By households and other cons.	113551	122699	119961	127041	132455	145359
Net installed capacity	Thousand kilowatts					
Total main activity and autoproducer	60737	66483	70356	74475	75816	*73001
Combustible fuels	52940	57896	60485	63596	64413	*63081
Hydro	7705	8488	8746	9746	10266	*8770
Nuclear	..	..	1020	1020	1020	*1020
Other	92	100	105	113	117	*130
Main activity	56506	61460	65222	68894	70235	*67420
Combustible fuels	48709	52872	55351	58015	58832	*57500
Hydro	7705	8488	8746	9746	10266	*8770
Nuclear	..	..	1020	1020	1020	*1020
Other	92	100	105	113	117	*130
Combustible fuel input	Terajoules					
Biogas	127	127	127	127	127	553
Gas-diesel oil	184649	221512	352009	290708	290783	321640
Fuel oil	366055	339889	461128	554397	585589	391961
Other kerosene	..	..	..	2935	6614	44
Natural gas	1693667	1751653	1517956	1587843	1430042	1974581
Coke-oven gas	106	0	0	0	35	202
Blast furnace gas	7336	7028	8480	7980	9088	7429
Total input	2251941	2320209	2339699	2443990	2322278	2696410
Total production	770082	803756	818741	863107	874472	921301
Estimated efficiency (% of production to input)	34	35	35	35	38	34

Statistics on electricity

Iraq

Item	2009	2010	2011	2012	2013	2014
Production, trade and consumption	**Million kilowatt-hours**					
Total main activity and autoproducer	46065	48909	54240	46018	58422	67768
Combustible fuels	42837	44142	50100	41626	53665	64837
Hydro	3228	4767	4140	4392	4757	2931
Nuclear	..	..	..	..	..	..
Other	..	..	..	..	..	..
Main activity	46065	48909	54240	46018	58422	67768
Combustible fuels	42837	44142	50100	41626	53665	64837
Hydro	3228	4767	4140	4392	4757	2931
Nuclear	..	..	..	..	..	..
Other	..	..	..	..	..	..
Own use in electricity, CHP and heat plants	2454	2411	0	1373	2310	1713
Net production	43611	46498	54240	44645	56112	66055
Imports	5604	6722	7262	10170	12202	12251
Exports	..	..	..	..	..	..
Losses	18425	*18425	18854	19797	23285	34312
Consumption	35973	40038	42648	35076	45042	43993
Energy industries own use	..	..	..	..	..	..
By industry and construction	5311	5567	8743	8244	9867	10071
By transport	..	..	..	..	..	..
By households and other cons.	30662	34471	33905	26832	35175	33922
Net installed capacity	**Thousand kilowatts**					
Total main activity and autoproducer	14384	14171	16952	16952	27110	25589
Combustible fuels	11871	11658	14439	14439	24597	23076
Hydro	2513	2513	2513	2513	2513	2513
Nuclear	..	..	..	..	..	..
Other	..	..	..	..	..	..
Main activity	14384	14171	16952	16952	27110	25589
Combustible fuels	11871	11658	14439	14439	24597	23076
Hydro	2513	2513	2513	2513	2513	2513
Nuclear	..	..	..	..	..	..
Other	..	..	..	..	..	..
Combustible fuel input	**Terajoules**					
Crude oil	144074	133710	132780	125039	191788	283114
Gas-diesel oil	..	..	217623	289089	333508	426345
Fuel oil	141723	162974	213837	282962	289506	229674
Naphtha	4806	3872	0	0	0	0
Natural gas	118486	144506	221803	218273	152218	183234
Total input	409089	445061	786043	915362	967021	1122367
Total production	154213	158911	180360	149854	193194	233413
Estimated efficiency (% of production to input)	38	36	23	16	20	21

Statistics on electricity

Ireland

Item	2009	2010	2011	2012	2013	2014
Production, trade and consumption	Million kilowatt-hours					
Total main activity and autoproducer	28313	28685	27474	27600	26142	26314
Combustible fuels	24101	25094	22386	22575	20655	20185
Hydro	1257	776	707	1014	944	988
Nuclear	..	..	..	..	..	..
Other	2955	2815	4381	4011	4543	5141
Main activity	26482	26738	25510	25474	24086	24241
Combustible fuels	22270	23147	20423	20450	18600	18113
Hydro	1257	776	707	1014	944	988
Nuclear	..	..	..	..	..	..
Other	2955	2815	4380	4010	4542	5140
Own use in electricity, CHP and heat plants	1186	1243	1110	1122	1025	1007
Net production	27127	27442	26364	26478	25117	25307
Imports	939	760	732	784	2625	2853
Exports	175	290	242	370	383	704
Losses	2101	2123	2052	2022	2034	2044
Consumption	26010	25923	25086	24733	24990	24814
Energy industries own use	742	503	215	570	789	678
By industry and construction	8598	9104	9484	9161	9287	9396
By transport	45	46	45	45	42	39
By households and other cons.	16625	16270	15342	14957	14872	14701
Net installed capacity	Thousand kilowatts					
Total main activity and autoproducer	7433	8311	8588	8589	8798	9078
Combustible fuels	5640	6407	6427	6295	6327	6337
Hydro	526	529	529	529	529	529
Nuclear	..	..	..	..	..	..
Other	1267	1375	1632	1765	1942	2212
Main activity	7131	8001	8260	8256	8460	8735
Combustible fuels	5339	6098	6100	5963	5990	5995
Hydro	526	529	529	529	529	529
Nuclear	..	..	..	..	..	..
Other	1266	1374	1631	1764	1941	2211
Combustible fuel input	Terajoules					
Hard Coal	36384	40909	39836	49285	42334	40473
Peat	28466	22386	21937	28176	22070	24120
Biogas	1994	2095	2014	1974	1721	1846
Gas-diesel oil	344	1075	344	301	172	301
Fuel oil	8282	4323	1697	1656	1414	1980
Refinery gas	248	297	248	347	198	198
Natural gas	128079	140430	115994	105362	97403	91594
Fuelwood	*558	*1020	*1276	*1718	*2141	*2437
Municipal waste	..	..	..	1799	2044	2096
Total input	204355	212535	183345	190619	169497	165045
Total production	86764	90338	80590	81270	74358	72666
Estimated efficiency (% of production to input)	42	43	44	43	44	44

Isle of Man

Item	2009	2010	2011	2012	2013	2014
Production, trade and consumption	**Million kilowatt-hours**					
Total main activity and autoproducer	533	498	499	296	*504	*504
Combustible fuels	529	495	496	292	*500	*500
Hydro	3	3	3	4	*4	*4
Nuclear	..	..	..	..	..	..
Other	..	..	..	..	..	..
Main activity	533	498	499	296	*504	*504
Combustible fuels	529	495	496	292	*500	*500
Hydro	3	3	3	4	*4	*4
Nuclear	..	..	..	..	..	..
Other	..	..	..	..	..	..
Own use in electricity, CHP and heat plants	0	0	0	0	*54	*43
Net production	533	498	499	296	*450	*461
Imports	19	45	41	180	*50	*50
Exports	114	108	115	48	77	93
Losses	48	48	47	47	*50	*50
Consumption	389	387	377	381	373	368
Energy industries own use	..	..	..	..	..	..
By industry and construction	133	137	133	134	137	137
By transport	..	..	..	..	..	..
By households and other cons.	256	250	244	247	236	231
Net installed capacity	**Thousand kilowatts**					
Total main activity and autoproducer	182	180	180	180	186	186
Combustible fuels	181	179	179	179	185	185
Hydro	1	1	1	1	1	1
Nuclear	..	..	..	..	..	..
Other	..	..	..	..	..	..
Main activity	182	180	180	180	186	186
Combustible fuels	181	179	179	179	185	185
Hydro	1	1	1	1	1	1
Nuclear	..	..	..	..	..	..
Other	..	..	..	..	..	..
Combustible fuel input	**Terajoules**					
Fuel oil	93	93	96	66	*121	*121
Natural gas	3569	3386	3375	1918	*3587	3587
Municipal waste	*449	*407	*429	*436	*440	*440
Total input	4111	3886	3900	2420	*4148	4148
Total production	1905	1782	1784	1049	*1800	*1800
Estimated efficiency (% of production to input)	46	46	46	43	43	43

Statistics on electricity

Israel

Item	2009	2010	2011	2012	2013	2014
Production, trade and consumption	Million kilowatt-hours					
Total main activity and autoproducer	55008	58591	59679	63003	61322	60813
Combustible fuels	54683	58450	59371	62595	60794	59954
Hydro	24	31	29	33	28	13
Nuclear	..	..	..	..	..	..
Other	301	110	279	375	500	846
Main activity	53192	56873	57855	61488	59026	57478
Combustible fuels	53155	56812	57819	61453	58997	57464
Hydro	20	27	24	29	23	8
Nuclear	..	..	..	..	..	..
Other	17	34	12	6	6	6
Own use in electricity, CHP and heat plants	4467	4875	4962	5252	2140	1989
Net production	50541	53716	54717	57751	59182	58824
Imports	..	..	..	..	..	..
Exports	3783	3966	4222	4434	4675	4844
Losses	1761	1616	1640	1736	2567	1738
Consumption	45443	49103	49512	52223	52118	52201
Energy industries own use	..	384	556	419	478	883
By industry and construction	10533	12488	12015	12448	13181	15211
By transport	..	..	..	..	..	..
By households and other cons.	34910	36231	36941	39356	38459	36107
Net installed capacity	Thousand kilowatts					
Total main activity and autoproducer	12264	13058	13990	14413	14988	16223
Combustible fuels	12166	12940	13662	14163	14491	15526
Hydro	7	7	7	7	10	10
Nuclear	..	..	..	..	..	..
Other	91	111	321	243	487	687
Main activity	11857	12583	13305	13746	14094	15078
Combustible fuels	11846	12547	13269	13735	14080	15064
Hydro	5	5	5	5	8	8
Nuclear	..	..	..	..	..	..
Other	6	31	31	6	6	6
Combustible fuel input	Terajoules					
Hard Coal	309572	308520	312358	351341	289694	273447
Brown Coal	0	0	3359	3199	3573	..
Biogas	115	347	391	777	442	638
Oil shale	1301	1266	1219	1240	1234	1161
Gas-diesel oil	8686	8815	27133	83893	13244	731
Fuel oil	14867	15110	13615	46339	6828	525
Other oil products	..	..	26009	24160	1930	1568
Natural gas	155310	193347	171628	76692	258879	263695
Fuelwood	*448	*367	*411	0	0	0
Total input	490300	527772	556124	587641	575822	541765
Total production	196859	210420	213736	225342	218858	215834
Estimated efficiency (% of production to input)	40	40	38	38	38	40

Statistics on electricity

Italy

Item	2009	2010	2011	2012	2013	2014
Production, trade and consumption	**Million kilowatt-hours**					
Total main activity and autoproducer	292640	302063	302582	299277	289807	279827
Combustible fuels	226033	230472	227711	216811	192237	175510
Hydro	53443	54406	47757	43854	54672	60256
Nuclear	..	..	..	..	..	..
Other	13164	17185	27114	38612	42898	44061
Main activity	272365	278197	279019	283220	273734	264085
Combustible fuels	207033	208049	205528	201786	177113	160761
Hydro	52554	53479	46889	43281	54045	59518
Nuclear	..	..	..	..	..	..
Other	12778	16669	26602	38153	42576	43806
Own use in electricity, CHP and heat plants	11534	11316	11141	11475	10974	10679
Net production	281106	290747	291441	287802	278833	269148
Imports	47070	45987	47519	45407	44338	46747
Exports	2111	1827	1787	2304	2200	3031
Losses	20352	20570	20848	21000	21187	19451
Consumption	305713	314336	316330	309911	299784	293414
Energy industries own use	15697	15023	14502	13169	12386	11916
By industry and construction	120607	127868	128076	120323	114981	112918
By transport	10535	10666	10793	10759	10775	10463
By households and other cons.	158874	160779	162959	165660	161642	158117
Net installed capacity	**Thousand kilowatts**					
Total main activity and autoproducer	101447	106488	118443	124234	124750	121762
Combustible fuels	73041	74658	75977	76793	74733	71272
Hydro	21371	21520	21737	21880	22009	22098
Nuclear	..	..	..	..	..	..
Other	7035	10310	20729	25561	28008	28392
Main activity	95333	100325	112667	120039	120448	117688
Combustible fuels	67231	68797	70501	72852	70664	67419
Hydro	21195	21348	21568	21752	21890	21979
Nuclear	..	..	..	..	..	..
Other	6907	10180	20598	25435	27894	28290
Combustible fuel input	**Terajoules**					
Hard Coal	400343	404623	455418	482749	427786	408150
Biogas	17863	21184	46073	47476	74133	80223
Fuel oil	155217	87628	68034	51510	41006	32401
Refinery gas	31730	34403	37472	36927	32274	26978
Other oil products	108821	167634	148619	139534	105163	106811
Natural gas	1229672	1295794	1225420	1112248	938888	829412
Blast furnace gas	22107	29591	34635	31494	23404	25051
Fuelwood	*39573	*35670	*55439	*59714	*78399	*82397
Municipal waste	57442	65178	70588	67560	69302	71882
Other liquid biofuels	8987	17892	16276	18796	22632	26386
Others	52903	48268	47806	49598	34860	30130
Total input	2124658	2207864	2205780	2097606	1847848	1719821
Total production	813719	829699	819760	780520	692053	631836
Estimated efficiency (% of production to input)	38	38	37	37	37	37

Statistics on electricity

Jamaica

Item	2009	2010	2011	2012	2013	2014
Production, trade and consumption	Million kilowatt-hours					
Total main activity and autoproducer	4208	4217	4171	4156	4256	4123
Combustible fuels	3999	4012	3926	3898	4017	3869
Hydro	158	152	152	151	124	135
Nuclear	..	..	..	..	..	..
Other	51	53	92	108	115	119
Main activity	2880	2807	2760	2615	2343	2462
Combustible fuels	2722	2656	2607	2464	2218	2327
Hydro	158	152	152	151	124	135
Nuclear	..	..	..	..	..	..
Other	..	..	..	..	..	..
Own use in electricity, CHP and heat plants	211	16	36	44	115	12
Net production	3996	4201	4135	4113	4141	4111
Imports	..	..	..	..	..	..
Exports	..	..	..	..	..	..
Losses	930	902	962	1033	1097	1103
Consumption	3067	3301	3176	3081	*3039	*3008
Energy industries own use	..	..	..	..	..	..
By industry and construction	1131	1059	1020	989	*984	*958
By transport	..	..	..	..	..	..
By households and other cons.	1935	2242	2156	2092	*2055	*2050
Net installed capacity	Thousand kilowatts					
Total main activity and autoproducer	815	819	840	912	938	945
Combustible fuels	772	776	776	848	874	874
Hydro	23	23	23	23	23	29
Nuclear	..	..	..	..	..	..
Other	21	21	42	42	42	42
Main activity	598	608	611	611	637	644
Combustible fuels	576	586	586	586	612	612
Hydro	23	23	23	23	23	29
Nuclear	..	..	..	..	..	..
Other	..	0	3	3	3	3
Combustible fuel input	Terajoules					
Gas-diesel oil	10523	10018	10226	9779	7860	8110
Fuel oil	29862	29862	29143	26281	26433	24433
Total input	40385	39880	39369	36060	34294	32543
Total production	14395	14442	14135	14031	14461	13928
Estimated efficiency (% of production to input)	36	36	36	39	42	43

105

Statistics on electricity

Japan

Item	2009	2010	2011	2012	2013	2014
Production, trade and consumption	**Million kilowatt-hours**					
Total main activity and autoproducer	1075305	1147899	1082230	1064065	1065623	1040676
Combustible fuels	702466	758578	876365	950187	950236	921613
Hydro	83832	90682	91709	83645	84923	86942
Nuclear	279750	288230	101761	15939	9303	0
Other	9257	10409	12395	14294	21161	32121
Main activity	925495	981949	916546	901063	897716	864204
Combustible fuels	568503	617017	737786	815167	817244	791407
Hydro	74539	74175	74378	67360	68601	70255
Nuclear	279750	288230	101761	15939	9303	0
Other	2703	2527	2621	2597	2568	2542
Own use in electricity, CHP and heat plants	40921	42375	37850	40737	37378	33797
Net production	1034384	1105524	1044380	1023328	1028245	1006879
Imports	..	..	..	..	..	..
Exports	..	..	..	..	..	..
Losses	48333	47149	46834	43244	47518	45414
Consumption	989230	1045412	988146	987133	990121	971729
Energy industries own use	21036	23847	23170	23651	21805	20242
By industry and construction	331328	336421	324349	313987	298171	295623
By transport	18815	18764	17667	17714	17861	17826
By households and other cons.	618051	666380	622960	631781	652284	638038
Net installed capacity	**Thousand kilowatts**					
Total main activity and autoproducer	284486	287027	292058	295193	302711	315318
Combustible fuels	183237	183882	186810	190405	192759	194857
Hydro	47243	47736	48418	48934	48932	49597
Nuclear	48847	48960	48960	46148	44264	44264
Other	5159	6449	7870	9706	16756	26600
Main activity	237773	227997	229908	231220	231467	234028
Combustible fuels	142574	135070	136132	139795	141901	143777
Hydro	45840	43367	44168	44652	44676	45403
Nuclear	48847	48960	48960	46148	44264	44264
Other	512	600	648	625	626	584
Combustible fuel input	**Terajoules**					
Hard Coal	2261840	2364118	2262611	2405761	2673352	2641246
Crude oil	131765	172119	418389	487381	418685	244621
Fuel oil	379841	386870	644057	779962	619090	466256
Refinery gas	61628	61677	59994	59301	70637	64301
Natural gas	2602151	2720318	3334450	3503837	3518496	3573682
Coke-oven gas	68405	77310	67201	70309	76269	71324
Blast furnace gas	185195	225927	209503	224011	237850	239530
Vegetal waste	0	*110429	*116418	*123311	*129642	*128546
Bitumen	56039	56923	53225	47396	51938	44903
Black liquor	*75480	*82658	*77620	*87230	*90572	*93282
Others	267767	226935	254467	259926	217185	206733
Total input	6090110	6485284	7497935	8048425	8103716	7774425
Total production	2528878	2730881	3154914	3420673	3420850	3317807
Estimated efficiency (% of production to input)	42	42	42	43	42	43

Statistics on electricity

Jersey

Item	2009	2010	2011	2012	2013	2014
Production, trade and consumption	**Million kilowatt-hours**					
Total main activity and autoproducer	64	56	46	60	328	135
Combustible fuels	64	56	46	60	328	135
Hydro	..	..	..	..	..	..
Nuclear	..	..	..	..	..	..
Other	..	..	..	..	..	..
Main activity	46	41	13	17	147	99
Combustible fuels	46	41	13	17	147	99
Hydro	..	..	..	..	..	..
Nuclear	..	..	..	..	..	..
Other	..	..	..	..	..	..
Own use in electricity, CHP and heat plants	0	0	0	17	22	20
Net production	64	56	46	43	306	115
Imports	661	669	700	681	553	581
Exports	..	..	..	..	..	..
Losses	74	77	67	68	49	45
Consumption	655	649	677	670	690	650
Energy industries own use	..	..	..	..	..	..
By industry and construction	286	285	297	298	285	285
By transport	..	..	..	..	..	..
By households and other cons.	369	364	380	372	405	364
Net installed capacity	**Thousand kilowatts**					
Total main activity and autoproducer	*217	*217	*219	*219	*240	*240
Combustible fuels	*217	*217	*219	*219	*240	*240
Hydro	..	..	..	..	..	..
Nuclear	..	..	..	..	..	..
Other	..	..	..	..	..	..
Main activity	*212	*212	*212	*212	*212	*212
Combustible fuels	*212	*212	*212	*212	*212	*212
Hydro	..	..	..	..	..	..
Nuclear	..	..	..	..	..	..
Other	..	..	..	..	..	..
Combustible fuel input	**Terajoules**					
Gas-diesel oil	*129	*106	146	189	1535	894
Fuel oil	*343	*408	*40	*40	*420	*166
Municipal waste	*430	*510	679	*686	*950	*700
Total input	*902	*1024	866	*916	2905	1760
Total production	230	203	165	217	1180	487
Estimated efficiency (% of production to input)	25	20	19	24	41	28

Statistics on electricity

Jordan

Item	2009	2010	2011	2012	2013	2014
Production, trade and consumption	**Million kilowatt-hours**					
Total main activity and autoproducer	14272	14777	14647	16596	17263	18270
Combustible fuels	14210	14713	14589	16532	17205	18210
Hydro	59	61	55	61	55	58
Nuclear	..	..	..	..	..	..
Other	3	3	3	3	3	2
Main activity	14009	14486	14390	16355	16977	17887
Combustible fuels	13947	14422	14332	16291	16919	17827
Hydro	59	61	55	61	55	58
Nuclear	..	..	..	..	..	..
Other	3	3	3	3	3	2
Own use in electricity, CHP and heat plants	536	544	616	619	616	632
Net production	13736	14233	14031	15977	16647	17638
Imports	383	670	1738	784	381	435
Exports	139	58	86	104	59	64
Losses	2024	1989	2148	2383	2406	2569
Consumption	12247	12919	13599	14352	14571	15419
Energy industries own use	99	82	103	107	99	98
By industry and construction	2907	3242	3446	3431	3425	3780
By transport	..	..	..	..	..	..
By households and other cons.	9241	9595	10050	10814	11047	11541
Net installed capacity	**Thousand kilowatts**					
Total main activity and autoproducer	2948	3178	3511	3511	3385	4199
Combustible fuels	2932	3162	3498	3498	3372	4186
Hydro	12	12	12	12	12	12
Nuclear	..	..	..	..	..	..
Other	4	4	1	1	1	1
Main activity	2749	2979	3312	3312	3186	4000
Combustible fuels	2733	2963	3299	3299	3173	3987
Hydro	12	12	12	12	12	12
Nuclear	..	..	..	..	..	..
Other	4	4	1	1	1	1
Combustible fuel input	**Terajoules**					
Biogas	84	108	96	72	72	72
Gas-diesel oil	654	3844	40755	64999	58961	70258
Fuel oil	13086	34728	52354	54649	53887	69561
Natural gas	143542	106470	40598	30652	37961	*12590
Total input	157365	145150	133804	150372	150881	152480
Total production	51156	52967	52520	59515	61938	65556
Estimated efficiency (% of production to input)	33	36	39	40	41	43

Kazakhstan

Item	2009	2010	2011	2012	2013	2014
Production, trade and consumption	Million kilowatt-hours					
Total main activity and autoproducer	78710	82646	86586	92817	103086	105068
Combustible fuels	71831	74624	78703	85177	95349	96791
Hydro	6879	8022	7883	7637	7731	8263
Nuclear	..	..	..		..	..
Other	..	..	..	3	6	14
Main activity	78710	82646	86586	92817	103085	105067
Combustible fuels	71831	74624	78703	85177	95349	96791
Hydro	6879	8022	7883	7637	7731	8263
Nuclear	..	..	..	..	..	..
Other	..	..	..	3	5	13
Own use in electricity, CHP and heat plants	15041	14788	10165	11867	17337	19778
Net production	63669	67858	76421	80950	85749	85290
Imports	1710	2105	3406	4252	2147	1749
Exports	2378	1560	1809	2933	2996	2917
Losses	6472	6612	6479	7142	11174	7082
Consumption	58142	63017	71296	75362	73338	76676
Energy industries own use	6257	5623	6544	7114	7611	7776
By industry and construction	35943	40121	44372	49162	44280	45529
By transport	2730	3124	3945	3352	2960	2511
By households and other cons.	13212	14149	16435	15734	18487	20860
Net installed capacity	Thousand kilowatts					
Total main activity and autoproducer	*18734	*19217	*20217	*21219	*22222	*25011
Combustible fuels	*16517	*17000	*18000	*19000	*20000	*22500
Hydro	*2217	*2217	*2217	*2217	*2217	*2500
Nuclear	..	..	..	..	..	..
Other	..	..	..	*2	*5	*11
Main activity	*18734	*19217	*20217	*21219	*22221	*25010
Combustible fuels	*16517	*17000	*18000	*19000	*20000	*22500
Hydro	*2217	*2217	*2217	*2217	*2217	*2500
Nuclear	..	..	..	..	..	..
Other	..	..	..	*2	*4	*10
Combustible fuel input	Terajoules					
Hard Coal	688277	726684	769012	817322	874273	877729
Brown Coal	24897	27784	24648	8924	615	630
Crude oil	85	..	..	..	..	..
Fuel oil	10625	7514	6585	7838	7151	12201
Natural gas	146934	83969	90739	171672	194836	233178
Total input	870818	845952	890984	1005756	1076875	1123738
Total production	258592	268646	283331	306637	343256	348448
Estimated efficiency (% of production to input)	30	32	32	30	32	31

Statistics on electricity

Kenya

Item	2009	2010	2011	2012	2013	2014
Production, trade and consumption	**Million kilowatt-hours**					
Total main activity and autoproducer	6507	6976	7560	7852	8448	9139
Combustible fuels	3047	2293	2882	2305	2218	2635
Hydro	2160	3224	3217	4016	4435	3569
Nuclear	..	..	..	..	..	..
Other	1300	1459	1461	1530	1796	2934
Main activity	6507	6976	7560	7852	8448	9139
Combustible fuels	3047	2293	2882	2305	2218	2635
Hydro	2160	3224	3217	4016	4435	3569
Nuclear	..	..	..	..	..	..
Other	1300	1459	1461	1530	1796	2934
Own use in electricity, CHP and heat plants	39	30	34	39	49	43
Net production	6468	6946	7526	7812	8399	9095
Imports	39	30	34	39	49	158
Exports	27	30	37	33	36	27
Losses	1052	1192	1249	1404	1476	1662
Consumption	5767	5729	6500	6760	7296	7786
Energy industries own use	..	..	..	..	..	..
By industry and construction	3375	3401	3636	3686	4438	3988
By transport	..	..	..	..	..	..
By households and other cons.	2392	2328	2864	3074	2858	3798
Net installed capacity	**Thousand kilowatts**					
Total main activity and autoproducer	1314	1418	1540	1611	1723	2094
Combustible fuels	424	495	609	637	715	734
Hydro	730	728	735	770	767	797
Nuclear	..	..	..	..	..	..
Other	160	195	197	204	242	563
Main activity	1314	1418	1540	1611	1723	2094
Combustible fuels	424	495	609	637	715	734
Hydro	730	728	735	770	767	797
Nuclear	..	..	..	..	..	..
Other	160	195	197	204	242	563
Combustible fuel input	**Terajoules**					
Gas-diesel oil	10320	10578	15308	12255	12943	7310
Fuel oil	13413	18140	15473	12403	6464	6399
Vegetal waste	16800	16800	*16800	*16800	16800	16800
Total input	40533	45518	47581	41458	36207	30509
Total production	10969	8255	10373	8299	7984	9487
Estimated efficiency (% of production to input)	27	18	22	20	22	31

Statistics on electricity

Kiribati

Item	2009	2010	2011	2012	2013	2014
Production, trade and consumption	Million kilowatt-hours					
Total main activity and autoproducer	*22	*24	*24	*24	*24	*24
Combustible fuels	*22	*24	*24	*24	*24	*24
Hydro	..	..	..	..	..	..
Nuclear	..	..	..	..	..	..
Other	..	..	..	..	..	..
Main activity	*22	*24	*24	*24	*24	*24
Combustible fuels	*22	*24	*24	*24	*24	*24
Hydro	..	..	..	..	..	..
Nuclear	..	..	..	..	..	..
Other	..	..	..	..	..	..
Own use in electricity, CHP and heat plants	*1	*1	*1	*1	*1	*1
Net production	*21	*23	*23	*23	*23	*23
Imports	..	..	..	..	..	..
Exports	..	..	..	..	..	..
Losses	..	..	..	..	..	..
Consumption	*21	*23	*23	*23	*23	*23
Energy industries own use	..	..	..	..	..	..
By industry and construction	..	..	..	..	..	..
By transport	..	..	..	..	..	..
By households and other cons.	*21	*23	*23	*23	*23	*23
Net installed capacity	Thousand kilowatts					
Total main activity and autoproducer	*6	*6	*6	*6	*6	*6
Combustible fuels	*5	*5	*5	*5	*5	*5
Hydro	..	..	..	..	..	..
Nuclear	..	..	..	..	..	..
Other	*1	*1	*1	*1	*1	*1
Main activity	*6	*6	*6	*6	*6	*6
Combustible fuels	*5	*5	*5	*5	*5	*5
Hydro	..	..	..	..	..	..
Nuclear	..	..	..	..	..	..
Other	*1	*1	*1	*1	*1	*1
Combustible fuel input	Terajoules					
Gas-diesel oil	258	282	*282	*282	*280	*288
Total input	258	282	*282	*282	*280	*288
Total production	*79	*87	*87	*87	*87	*87
Estimated efficiency (% of production to input)	31	31	31	31	31	30

111

Statistics on electricity

Korea, Dem.Ppl's.Rep.

Item	2009	2010	2011	2012	2013	2014
Production, trade and consumption	**Million kilowatt-hours**					
Total main activity and autoproducer	21127	21664	19226	19236	18332	17909
Combustible fuels	8627	8264	6026	5736	4432	4909
Hydro	12500	13400	13200	13500	13900	13000
Nuclear	..	..	..	..	..	..
Other	..	..	..	..	..	..
Main activity	21127	21664	19226	19236	18332	17909
Combustible fuels	8627	8264	6026	5736	4432	4909
Hydro	12500	13400	13200	13500	13900	13000
Nuclear	..	..	..	..	..	..
Other	..	..	..	..	..	..
Own use in electricity, CHP and heat plants	2004	2055	1824	1825	1739	1699
Net production	19123	19609	17402	17411	16593	16210
Imports	..	..	..	..	..	..
Exports	..	..	..	..	..	..
Losses	3340	3425	3040	3042	2899	2832
Consumption	15783	16184	14362	14369	13694	13378
Energy industries own use	..	..	..	..	..	..
By industry and construction	7892	8092	7181	7185	6847	6689
By transport	..	..	..	..	..	..
By households and other cons.	7891	8092	7181	7184	6847	6689
Net installed capacity	**Thousand kilowatts**					
Total main activity and autoproducer	*9500	*9500	*9500	*9500	*9500	*9500
Combustible fuels	*4500	*4500	*4500	*4500	*4500	*4500
Hydro	*5000	*5000	*5000	*5000	*5000	*5000
Nuclear	..	..	..	..	..	..
Other	..	..	..	..	..	..
Main activity	*9500	*9500	*9500	*9500	*9500	*9500
Combustible fuels	*4500	*4500	*4500	*4500	*4500	*4500
Hydro	*5000	*5000	*5000	*5000	*5000	*5000
Nuclear	..	..	..	..	..	..
Other	..	..	..	..	..	..
Combustible fuel input	**Terajoules**					
Hard Coal	97225	90351	62580	63681	43377	50395
Brown Coal	11448	10973	7720	7403	5504	6155
Fuel oil	10787	10302	11150	9979	10342	10625
Total input	119460	111626	81450	81063	59223	67175
Total production	31057	29750	21694	20650	15955	17672
Estimated efficiency (% of production to input)	26	27	27	25	27	26

Statistics on electricity

Korea, Republic of

Item	2009	2010	2011	2012	2013	2014
Production, trade and consumption	**Million kilowatt-hours**					
Total main activity and autoproducer	454504	499508	523286	534618	541996	550933
Combustible fuels	299668	342502	358505	374116	390858	381510
Hydro	5641	6472	7831	7652	8394	7820
Nuclear	147771	148596	154723	150327	138784	156407
Other	1424	1938	2227	2523	3960	5196
Main activity	430618	471026	493596	503660	513833	517962
Combustible fuels	275939	314284	329200	343479	363146	349044
Hydro	5641	6472	7831	7652	8394	7820
Nuclear	147771	148596	154723	150327	138784	156407
Other	1267	1674	1842	2202	3509	4691
Own use in electricity, CHP and heat plants	19526	19116	20136	20272	20526	20926
Net production	434978	480392	503150	514346	521470	530007
Imports	..	..	..	..	..	..
Exports	..	..	..	..	..	..
Losses	16770	18034	17430	17292	18311	18270
Consumption	418403	462136	485215	497357	503570	505690
Energy industries own use	12704	12791	14602	15952	16445	18856
By industry and construction	199326	228120	246046	250643	255697	259606
By transport	2174	2191	2246	2250	2168	2003
By households and other cons.	204199	219034	222321	228512	229260	225225
Net installed capacity	**Thousand kilowatts**					
Total main activity and autoproducer	80611	84700	84654	87830	91485	99830
Combustible fuels	56482	60389	58303	59086	61795	69122
Hydro	5515	5525	6418	6447	6454	6467
Nuclear	17716	17716	18716	20716	20716	20716
Other	898	1070	1217	1581	2520	3525
Main activity	74722	79102	78952	81013	85147	92115
Combustible fuels	50706	54913	52800	52634	55960	62121
Hydro	5515	5525	6418	6447	6454	6467
Nuclear	17716	17716	18716	20716	20716	20716
Other	785	948	1018	1216	2017	2811
Combustible fuel input	**Terajoules**					
Hard Coal	1891050	2005875	2123995	2138543	2135854	2049054
Brown Coal	74588	106554	88237	45119	38499	103882
Fuel oil	126088	114736	84274	132754	139663	66741
Naphtha	58740	73025	76051	81613	46903	47882
Petroleum coke	..	715	1625	4128	8840	11050
Other oil products	..	..	3980	2774	3618	9929
Natural gas	572789	831449	928420	1027978	1054747	933846
Coke-oven gas	11010	24788	31841	22352	23544	38166
Blast furnace gas	78103	110292	135788	119240	102556	127625
Other recovered gases	8550	11795	15145	15976	14983	20893
Others	19175	22408	35853	36088	32147	36356
Total input	2840093	3301637	3525209	3626564	3601355	3445425
Total production	1078805	1233007	1290618	1346818	1407089	1373436
Estimated efficiency (% of production to input)	38	37	37	37	39	40

Statistics on electricity

Kuwait

Item	2009	2010	2011	2012	2013	2014
Production, trade and consumption	Million kilowatt-hours					
Total main activity and autoproducer	53216	57029	57457	62655	60982	65140
Combustible fuels	53216	57029	57457	62655	60982	65140
Hydro	..	..	..	..	..	..
Nuclear	..	..	..	..	..	..
Other	..	..	..	..	..	..
Main activity	53216	57029	57457	62655	60982	65140
Combustible fuels	53216	57029	57457	62655	60982	65140
Hydro	..	..	..	..	..	..
Nuclear	..	..	..	..	..	..
Other	..	..	..	..	..	..
Own use in electricity, CHP and heat plants	6615	6896	6948	7577	7375	7587
Net production	46601	50133	50509	55078	53607	57553
Imports	..	..	..	..	..	..
Exports	..	..	..	..	..	..
Losses	6615	6893	7082	8893	7398	7596
Consumption	39986	43240	43427	46185	46209	49957
Energy industries own use	5592	6016	6045	6451	6430	6905
By industry and construction	..	..	..	..	..	..
By transport	..	..	..	..	..	..
By households and other cons.	34394	37224	37382	39734	39779	43052
Net installed capacity	Thousand kilowatts					
Total main activity and autoproducer	12679	12679	14803	14803	15719	15819
Combustible fuels	12679	12679	14803	14803	15719	15819
Hydro	..	..	..	..	..	..
Nuclear	..	..	..	..	..	..
Other	..	..	..	..	..	..
Main activity	12579	12579	14703	14703	15619	15719
Combustible fuels	12579	12579	14703	14703	15619	15719
Hydro	..	..	..	..	..	..
Nuclear	..	..	..	..	..	..
Other	..	..	..	..	..	..
Combustible fuel input	Terajoules					
Crude oil	116113	103804	112560	95725	71825	69795
Gas-diesel oil	41237	52374	52761	57534	55986	59813
Fuel oil	347804	266317	268296	292577	284780	304212
Natural gas	165879	225651	249590	259312	256730	251248
Total input	671033	648146	683208	705148	669321	685068
Total production	191578	205304	206845	225558	219535	234504
Estimated efficiency (% of production to input)	29	32	30	32	33	34

Kyrgyzstan

Item	2009	2010	2011	2012	2013	2014
Production, trade and consumption	Million kilowatt-hours					
Total main activity and autoproducer	11083	12100	15158	15168	14011	14572
Combustible fuels	866	992	1019	989	914	1274
Hydro	10217	11108	14139	14179	13097	13298
Nuclear	..	..	..	..	..	..
Other	..	..	..	..	..	..
Main activity	11083	12100	15158	15168	14004	14552
Combustible fuels	866	992	1019	989	907	1254
Hydro	10217	11108	14139	14179	13097	13298
Nuclear	..	..	..	..	..	..
Other	..	..	..	..	..	..
Own use in electricity, CHP and heat plants	229	238	233	235	199	270
Net production	10854	11862	14925	14933	13812	14302
Imports	52	116	174	177	..	286
Exports	1251	1828	2848	1840	377	72
Losses	2758	2915	3389	3361	2841	3458
Consumption	6888	7121	8766	9661	10322	11058
Energy industries own use	36	43	54	70	55	74
By industry and construction	1893	1764	2000	1985	1464	1175
By transport	..	..	..	..	41	232
By households and other cons.	4959	5314	6712	7606	8762	9577
Net installed capacity	Thousand kilowatts					
Total main activity and autoproducer	3740	3859	*3859	*3859	*3861	*3864
Combustible fuels	796	795	*795	*795	*797	*800
Hydro	2944	3064	*3064	*3064	*3064	*3064
Nuclear	..	..	..	..	..	..
Other	..	..	..	..	..	..
Main activity	3740	3859	*3859	*3859	*3859	*3859
Combustible fuels	796	795	*795	*795	*795	*795
Hydro	2944	3064	*3064	*3064	*3064	*3064
Nuclear	..	..	..	..	..	..
Other	..	..	..	..	..	..
Combustible fuel input	Terajoules					
Hard Coal	16834	13750	15088	16240	9840	11838
Brown Coal	..	..	..	..	4645	7298
Gas-diesel oil	43	43	43	0	..	..
Fuel oil	4888	2464	2747	2222	1252	1050
Natural gas	3211	3621	2521	3254	382	1651
Total input	24977	19878	20399	21716	16120	21837
Total production	3118	3571	3668	3560	3290	4586
Estimated efficiency (% of production to input)	12	18	18	16	20	21

Statistics on electricity

Lao People's Dem. Rep.

Item	2009	2010	2011	2012	2013	2014
Production, trade and consumption	**Million kilowatt-hours**					
Total main activity and autoproducer	3874	3755	3802	*3802	4471	4380
Combustible fuels	*300	*300	*300	*300	*300	*300
Hydro	3381	3262	3309	*3309	3976	3885
Nuclear	..	..	..	..	..	..
Other	*193	*193	*193	*193	*195	*195
Main activity	3681	3562	3609	*3609	4276	4185
Combustible fuels	*300	*300	*300	*300	*300	*300
Hydro	3381	3262	3309	*3309	3976	3885
Nuclear	..	..	..	..	..	..
Other	..	..	..	..	..	..
Own use in electricity, CHP and heat plants	0	0	0	0	495	495
Net production	3874	3755	3802	*3802	3976	3885
Imports	819	999	748	*748	1205	1486
Exports	1939	2020	*2020	*2020	876	630
Losses	*198	*167	*211	*211	405	400
Consumption	1901	2228	2399	2853	3363	3768
Energy industries own use	..	..	..	..	..	..
By industry and construction	404	495	584	681	1118	1564
By transport	..	..	..	..	..	..
By households and other cons.	1498	1734	1815	2172	2245	2204
Net installed capacity	**Thousand kilowatts**					
Total main activity and autoproducer	2031	2746	2796	3199	3206	3284
Combustible fuels	*50	*50	*50	*50	*50	*50
Hydro	1805	2520	2570	2973	2980	3058
Nuclear	..	..	..	..	..	..
Other	*176	*176	*176	*176	*176	*176
Main activity	1855	2570	2620	3023	3030	3108
Combustible fuels	*50	*50	*50	*50	*50	*50
Hydro	1805	2520	2570	2973	2980	3058
Nuclear	..	..	..	..	..	..
Other	..	..	..	..	..	..
Combustible fuel input	**Terajoules**					
Hard Coal	*2451	*2457	*2457	*2457	*2451	*2503
Fuel oil	*889	*889	*889	*889	*889	*889
Total input	*3340	*3346	*3346	*3346	*3340	*3391
Total production	*1080	*1080	*1080	*1080	*1080	*1080
Estimated efficiency (% of production to input)	32	32	32	32	32	32

Latvia

Item	2009	2010	2011	2012	2013	2014
Production, trade and consumption	**Million kilowatt-hours**					
Total main activity and autoproducer	5569	6627	6095	6167	6209	5141
Combustible fuels	2062	3058	3137	2346	3177	3006
Hydro	3457	3520	2887	3707	2912	1994
Nuclear	..	..	..	..	..	..
Other	50	49	71	114	120	141
Main activity	5494	6544	6007	5984	5952	4884
Combustible fuels	1999	2990	3058	2170	2926	2760
Hydro	3449	3510	2878	3700	2909	1987
Nuclear	..	..	..	..	..	..
Other	46	44	71	114	117	137
Own use in electricity, CHP and heat plants	378	558	531	448	409	409
Net production	5191	6069	5564	5719	5800	4732
Imports	4259	3973	4009	4935	5005	5340
Exports	2605	3100	2764	3244	3650	3023
Losses	741	725	616	559	575	465
Consumption	6103	6215	6191	6848	6576	6582
Energy industries own use	..	..	..	..	..	..
By industry and construction	1506	1590	1670	1993	1808	1667
By transport	121	126	124	129	124	117
By households and other cons.	4476	4499	4397	4726	4644	4798
Net installed capacity	**Thousand kilowatts**					
Total main activity and autoproducer	2526	2580	2604	2702	2963	2967
Combustible fuels	953	967	986	1062	1305	1306
Hydro	1541	1581	1581	1579	1591	1592
Nuclear	..	..	..	..	..	..
Other	32	32	37	61	67	69
Main activity	2502	2557	2577	2660	2911	2924
Combustible fuels	937	951	965	1023	1255	1265
Hydro	1536	1576	1576	1576	1589	1590
Nuclear	..	..	..	..	..	..
Other	29	30	36	61	67	69
Combustible fuel input	**Terajoules**					
Hard Coal	446	393	393	499	446	169
Peat	0	0	0	0	40	0
Biogas	310	439	768	1890	2365	2780
Gas-diesel oil	0	0	0	0	0	0
Fuel oil	323	202	202	162	121	0
Other oil products	0	0	0	0	0	0
Natural gas	26708	34695	32210	28994	33458	29896
Fuelwood	703	813	877	1946	5618	7646
Vegetal waste	0	1	0	0	0	0
Biodiesel	0	0	37	37	0	0
Total input	28490	36543	34487	33527	42049	40491
Total production	7423	11009	11293	8446	11437	10822
Estimated efficiency (% of production to input)	26	30	33	25	27	27

Statistics on electricity

Lebanon

Item	2009	2010	2011	2012	2013	2014
Production, trade and consumption	**Million kilowatt-hours**					
Total main activity and autoproducer	13771	15712	16365	14826	17487	17952
Combustible fuels	13149	14873	15560	13819	16218	17759
Hydro	622	839	805	1007	1269	193
Nuclear	..	..	..	..	..	..
Other	..	..	..	..	..	..
Main activity	10810	11212	11615	8826	10487	12387
Combustible fuels	10188	10373	10810	7819	9218	12194
Hydro	622	839	805	1007	1269	193
Nuclear	..	..	..	..	..	..
Other	..	..	..	..	..	..
Own use in electricity, CHP and heat plants	0	0	0	0	0	0
Net production	13771	15712	16365	14826	17487	17952
Imports	1155	1245	840	323	522	136
Exports	..	..	..	..	..	..
Losses	1789	1869	1868	1372	1651	1879
Consumption	13137	15088	15337	13776	16358	16210
Energy industries own use	..	..	..	..	..	..
By industry and construction	3448	3961	4026	3616	4294	4255
By transport	..	..	..	..	..	..
By households and other cons.	9689	11127	11311	10160	12064	11955
Net installed capacity	**Thousand kilowatts**					
Total main activity and autoproducer	2908	3212	3262	3262	3262	3262
Combustible fuels	2634	2938	2988	2988	2988	2988
Hydro	274	*274	*274	*274	*274	*274
Nuclear	..	..	..	..	..	..
Other	..	..	..	..	..	..
Main activity	2308	2312	2312	2312	2312	2312
Combustible fuels	2034	2038	2038	2038	2038	2038
Hydro	274	*274	*274	*274	*274	*274
Nuclear	..	..	..	..	..	..
Other	..	..	..	..	..	..
Combustible fuel input	**Terajoules**					
Gas-diesel oil	77701	86774	102641	115455	118981	111800
Fuel oil	49611	51833	47955	39915	47793	53207
Natural gas	1931	9867	0	0	0	0
Total input	129243	148474	150596	155370	166774	165007
Total production	47336	53543	56016	49748	58385	63932
Estimated efficiency (% of production to input)	37	36	37	32	35	39

118

Statistics on electricity

Lesotho

Item	2009	2010	2011	2012	2013	2014
Production, trade and consumption	Million kilowatt-hours					
Total main activity and autoproducer	505	501	490	485	515	515
Combustible fuels	..	..	..	..	..	..
Hydro	505	501	490	485	515	515
Nuclear	..	..	..	..	..	..
Other	..	..	..	..	..	..
Main activity	505	501	490	485	515	515
Combustible fuels	..	..	..	..	..	..
Hydro	505	501	490	485	515	515
Nuclear	..	..	..	..	..	..
Other	..	..	..	..	..	..
Own use in electricity, CHP and heat plants	0	0	0	0	0	0
Net production	505	501	490	485	515	515
Imports	124	146	262	310	285	271
Exports	2	6	39	21	2	3
Losses	92	72	67	98	88	98
Consumption	536	568	645	676	711	*685
Energy industries own use	..	..	..	..	..	..
By industry and construction	209	209	221	232	236	*225
By transport	..	..	..	..	..	..
By households and other cons.	327	360	425	444	475	*460
Net installed capacity	Thousand kilowatts					
Total main activity and autoproducer	80	80	80	80	80	80
Combustible fuels	..	..	..	..	..	..
Hydro	80	80	80	80	80	80
Nuclear	..	..	..	..	..	..
Other	..	..	..	..	..	..
Main activity	80	80	80	80	80	80
Combustible fuels	..	..	..	..	..	..
Hydro	80	80	80	80	80	80
Nuclear	..	..	..	..	..	..
Other	..	..	..	..	..	..
Combustible fuel input	Terajoules					
Total input	..	..	..	..	..	..
Total production	..	..	..	..	..	..
Estimated efficiency (% of production to input)	..	..	..	..	..	..

Statistics on electricity

Liberia

Item	2009	2010	2011	2012	2013	2014
Production, trade and consumption	Million kilowatt-hours					
Total main activity and autoproducer	*250	*250	*300	*300	*300	*300
Combustible fuels	*250	*250	*300	*300	*300	*300
Hydro	..	..	..	..	..	..
Nuclear	..	..	..	..	..	..
Other	..	..	..	..	..	..
Main activity	*50	*50	*100	*100	*100	*100
Combustible fuels	*50	*50	*100	*100	*100	*100
Hydro	..	..	..	..	..	..
Nuclear	..	..	..	..	..	..
Other	..	..	..	..	..	..
Own use in electricity, CHP and heat plants	*20	*20	*20	*20	*20	*20
Net production	*230	*230	*280	*280	*280	*280
Imports	..	..	..	..	..	..
Exports	..	..	..	..	..	..
Losses	*15	*15	*24	*24	*24	*24
Consumption	*215	*215	*256	*256	*256	*256
Energy industries own use	..	..	..	..	..	..
By industry and construction	..	..	..	..	..	..
By transport	..	..	..	..	..	..
By households and other cons.	*215	*215	*256	*256	*256	*256
Net installed capacity	Thousand kilowatts					
Total main activity and autoproducer	*78	*78	*91	*91	*91	*91
Combustible fuels	*78	*78	*91	*91	*91	*91
Hydro	..	..	..	..	..	..
Nuclear	..	..	..	..	..	..
Other	..	..	..	..	..	..
Main activity	*8	*8	*21	*21	*21	*21
Combustible fuels	*8	*8	*21	*21	*21	*21
Hydro	..	..	..	..	..	..
Nuclear	..	..	..	..	..	..
Other	..	..	..	..	..	..
Combustible fuel input	Terajoules					
Gas-diesel oil	*1310	*1946	*2137	*2620	*2620	*2633
Fuel oil	*1616	*1212	*1212	*1212	*1212	*1212
Total input	*2926	*3158	*3349	*3832	*3832	*3845
Total production	*900	*900	*1080	*1080	*1080	*1080
Estimated efficiency (% of production to input)	31	28	32	28	28	28

Statistics on electricity

Libya

Item	2009	2010	2011	2012	2013	2014
Production, trade and consumption	Million kilowatt-hours					
Total main activity and autoproducer	30373	32558	25999	34275	37913	37731
Combustible fuels	30373	32558	25999	34275	37913	37731
Hydro	..	..	..	..	..	..
Nuclear	..	..	..	..	..	..
Other	..	..	..	..	..	..
Main activity	30373	32558	25999	34275	37913	37731
Combustible fuels	30373	32558	25999	34275	37913	37731
Hydro	..	..	..	..	..	..
Nuclear	..	..	..	..	..	..
Other	..	..	..	..	..	..
Own use in electricity, CHP and heat plants	950	1039	1929	1094	601	632
Net production	29423	31519	24070	33181	37312	37099
Imports	95	70	56	61	64	88
Exports	124	152	121	14	1	..
Losses	11069	11641	8197	20234	23309	26297
Consumption	18325	19796	15808	12994	14066	10890
Energy industries own use	..	..	..	..	..	..
By industry and construction	1934	2079	1660	1449	1892	1560
By transport	..	..	..	..	..	..
By households and other cons.	16391	17717	14148	11545	12174	9330
Net installed capacity	Thousand kilowatts					
Total main activity and autoproducer	6766	7066	8907	8907	9455	9455
Combustible fuels	6766	7066	8907	8907	9455	9455
Hydro	..	..	..	..	..	..
Nuclear	..	..	..	..	..	..
Other	..	..	..	..	..	..
Main activity	6766	7066	8907	8907	9455	9455
Combustible fuels	6766	7066	8907	8907	9455	9455
Hydro	..	..	..	..	..	..
Nuclear	..	..	..	..	..	..
Other	..	..	..	..	..	..
Combustible fuel input	Terajoules					
Gas-diesel oil	131494	108317	64973	86602	134246	143104
Fuel oil	77245	81042	59469	77083	36602	42703
Natural gas	125329	161516	163688	211555	236429	216420
Total input	334068	350875	288130	375240	407277	402227
Total production	109343	117209	93596	123390	136487	135832
Estimated efficiency (% of production to input)	33	33	32	33	34	34

Statistics on electricity

Liechtenstein

Item	2009	2010	2011	2012	2013	2014
Production, trade and consumption	**Million kilowatt-hours**					
Total main activity and autoproducer	71	79	72	86	85	40
Combustible fuels	4	4	4	4	3	3
Hydro	66	72	62	74	69	22
Nuclear	..	..	..	..	..	..
Other	1	3	6	9	13	16
Main activity	71	79	72	86	85	40
Combustible fuels	4	4	4	4	3	3
Hydro	66	72	62	74	69	22
Nuclear	..	..	..	..	..	..
Other	1	3	6	9	13	16
Own use in electricity, CHP and heat plants	0	0	0	0	0	0
Net production	71	79	72	86	85	40
Imports	307	318	326	318	319	355
Exports	..	..	..	..	..	..
Losses	..	..	..	..	..	..
Consumption	*378	*397	*398	*404	404	395
Energy industries own use	..	..	..	..	..	..
By industry and construction	..	..	..	..	..	..
By transport	..	..	..	..	..	..
By households and other cons.	*378	*397	*398	*404	404	395
Net installed capacity	**Thousand kilowatts**					
Total main activity and autoproducer	*12	*12	*12	*12	*13	*13
Combustible fuels	*1	*1	*1	*1	*1	1
Hydro	*10	*10	*10	*10	*10	*10
Nuclear	..	..	..	..	..	..
Other	*1	*1	*1	*1	*2	*2
Main activity	*12	*12	*12	*12	*13	*13
Combustible fuels	*1	*1	*1	*1	*1	1
Hydro	*10	*10	*10	*10	*10	*10
Nuclear	..	..	..	..	..	..
Other	*1	*1	*1	*1	*2	*2
Combustible fuel input	**Terajoules**					
Total input	..	..	..	..	..	..
Total production	15	15	14	13	12	9
Estimated efficiency (% of production to input)	..	..	..	..	..	..

122

Statistics on electricity

Lithuania

Item	2009	2010	2011	2012	2013	2014
Production, trade and consumption	Million kilowatt-hours					
Total main activity and autoproducer	15358	5749	4822	5043	4762	4397
Combustible fuels	2938	3980	3034	3337	2808	2354
Hydro	1139	1295	1056	937	1069	1088
Nuclear	10852	..	..	..	..	..
Other	429	474	732	769	885	955
Main activity	14736	5197	4222	4256	4141	3661
Combustible fuels	2587	3678	2691	2777	2424	1861
Hydro	1139	1295	1056	937	1069	1088
Nuclear	10852	..	..	..	..	..
Other	158	224	475	542	648	712
Own use in electricity, CHP and heat plants	1211	402	377	354	311	253
Net production	14147	5347	4445	4689	4451	4144
Imports	4783	8174	8086	8060	8073	8521
Exports	7715	2184	1347	1441	1127	898
Losses	969	989	872	883	872	815
Consumption	10240	10344	10308	10422	10522	10949
Energy industries own use	1869	2012	1728	1501	1567	1712
By industry and construction	2430	2654	2768	2893	2986	3119
By transport	77	76	74	75	74	63
By households and other cons.	5864	5602	5738	5953	5895	6055
Net installed capacity	Thousand kilowatts					
Total main activity and autoproducer	4909	3768	3940	4484	4568	4285
Combustible fuels	2727	2734	2837	3301	3320	3026
Hydro	876	876	876	876	876	877
Nuclear	1183	..	..	..	..	..
Other	123	158	227	307	372	382
Main activity	4689	3545	3666	4212	4298	4012
Combustible fuels	2532	2536	2588	3054	3075	2778
Hydro	876	876	876	876	876	877
Nuclear	1183	..	..	..	..	..
Other	98	133	202	282	347	357
Combustible fuel input	Terajoules					
Peat	0	0	0	0	0	35
Biogas	108	229	335	353	473	603
Gas-diesel oil	0	0	0	0	0	0
Fuel oil	6020	5292	1980	6141	2666	1414
Liquefied petroleum gas	0	0	0	0	0	0
Refinery gas	0	0	0	198	149	99
Natural gas	39295	51946	42383	37286	30737	24187
Fuelwood	*2331	*2472	*2359	*3786	*6075	6060
Municipal waste	..	..	..	..	929	975
Industrial waste	..	..	..	..	155	258
Others	0	0	0	0	0	0
Total input	47754	59939	47057	47764	41184	33631
Total production	10577	14328	10922	12013	10109	8474
Estimated efficiency (% of production to input)	22	24	23	25	25	25

Luxembourg

Item	2009	2010	2011	2012	2013	2014
Production, trade and consumption	**Million kilowatt-hours**					
Total main activity and autoproducer	3879	4590	3718	3818	2888	2967
Combustible fuels	2963	3046	2498	2543	1573	1623
Hydro	833	1468	1130	1160	1158	1169
Nuclear	..	..	..	..	..	..
Other	83	76	90	115	157	175
Main activity	3672	4374	3510	3593	2623	2696
Combustible fuels	2782	2859	2320	2361	1388	1453
Hydro	827	1460	1126	1155	1152	1163
Nuclear	..	..	..	..	..	..
Other	63	55	64	77	83	80
Own use in electricity, CHP and heat plants	47	31	20	34	29	29
Net production	3832	4559	3698	3784	2859	2938
Imports	6022	7280	7096	6732	6852	6961
Exports	2604	3216	2614	2622	1907	2067
Losses	116	122	121	121	121	120
Consumption	7139	8518	8062	7785	7570	7724
Energy industries own use	1025	1913	1533	1515	1468	1496
By industry and construction	3099	3633	3605	3227	3045	3175
By transport	112	120	129	129	129	123
By households and other cons.	2903	2852	2795	2914	2928	2930
Net installed capacity	**Thousand kilowatts**					
Total main activity and autoproducer	1704	1711	1741	1789	1813	2022
Combustible fuels	501	504	521	522	526	524
Hydro	1134	1134	1134	1134	1134	1330
Nuclear	..	..	..	..	..	..
Other	69	73	86	133	153	168
Main activity	1616	1619	1640	1653	1657	1851
Combustible fuels	441	443	463	463	467	465
Hydro	1132	1132	1132	1132	1132	1328
Nuclear	..	..	..	..	..	..
Other	43	44	45	58	58	58
Combustible fuel input	**Terajoules**					
Biogas	199	275	282	305	307	384
Gas-diesel oil	43	0	43	43	43	0
Natural gas	23027	24176	19997	20548	13490	13871
Fuelwood	0	0	0	0	*49	*525
Municipal waste	1224	1113	1225	1203	1185	1146
Total input	24493	25564	21547	22099	15074	15926
Total production	10667	10966	8993	9155	5663	5843
Estimated efficiency (% of production to input)	44	43	42	41	38	37

Madagascar

Item	2009	2010	2011	2012	2013	2014
Production, trade and consumption	**Million kilowatt-hours**					
Total main activity and autoproducer	1274	1360	1438	1520	1594	1659
Combustible fuels	533	649	747	765	784	774
Hydro	741	711	690	755	809	884
Nuclear	..	..	..	..	..	..
Other	0	0	0	0	*1	*1
Main activity	1104	1190	1268	1350	1424	1489
Combustible fuels	363	479	577	595	614	604
Hydro	741	711	690	755	809	884
Nuclear	..	..	..	..	..	..
Other	0	0	0	0	*1	*1
Own use in electricity, CHP and heat plants	18	19	21	23	*24	25
Net production	1256	1340	1417	1497	1570	1634
Imports	..	..	..	..	..	..
Exports	..	..	..	..	..	..
Losses	40	*40	*41	*41	*41	*42
Consumption	*961	1014	1053	1100	1222	1266
Energy industries own use	..	..	..	..	..	..
By industry and construction	*481	*489	*501	*516	611	633
By transport	..	..	..	..	..	..
By households and other cons.	480	525	552	584	611	633
Net installed capacity	**Thousand kilowatts**					
Total main activity and autoproducer	431	507	*507	*507	*508	*508
Combustible fuels	282	379	*379	*379	*379	*379
Hydro	148	128	*128	*128	*128	*128
Nuclear	..	..	..	..	..	..
Other	0	0	0	0	*1	*1
Main activity	399	475	*475	*475	*476	*476
Combustible fuels	250	347	*347	*347	*347	*347
Hydro	148	128	*128	*128	*128	*128
Nuclear	..	..	..	..	..	..
Other	0	0	0	0	*1	*1
Combustible fuel input	**Terajoules**					
Gas-diesel oil	*3966	*5061	*5061	*5061	*5160	*5160
Fuel oil	1843	*2022	*2022	*2022	*2101	*2141
Total input	*5809	*7082	*7082	*7082	*7261	*7301
Total production	1918	2336	2690	2753	2824	2786
Estimated efficiency (% of production to input)	33	33	38	39	39	38

Statistics on electricity

Malawi

Item	2009	2010	2011	2012	2013	2014
Production, trade and consumption	**Million kilowatt-hours**					
Total main activity and autoproducer	1944	2024	1979	2027	2098	*2098
Combustible fuels	*181	*181	*181	*181	*181	*181
Hydro	1763	1843	1798	1846	1917	*1917
Nuclear	..	..	..	..	..	..
Other	..	..	..	..	..	..
Main activity	1764	1844	1798	1846	1917	*1917
Combustible fuels	*5	*5	*5	*5	*5	*5
Hydro	1759	1839	1793	1841	1912	*1912
Nuclear	..	..	..	..	..	..
Other	..	..	..	..	..	..
Own use in electricity, CHP and heat plants	*91	*97	*95	*101	*103	*103
Net production	1853	1927	1884	1926	1995	*1995
Imports	..	..	..	..	..	..
Exports	16	21	20	18	*18	*18
Losses	*184	*191	*186	*191	*198	*198
Consumption	1652	1716	1678	1717	*1779	*1779
Energy industries own use	..	..	..	..	..	..
By industry and construction	553	556	617	503	*550	*570
By transport	..	..	..	..	..	..
By households and other cons.	1099	1160	1062	1214	*1229	*1209
Net installed capacity	**Thousand kilowatts**					
Total main activity and autoproducer	502	502	502	502	502	501
Combustible fuels	*211	*211	*211	*211	*211	*211
Hydro	291	291	291	291	291	290
Nuclear	..	..	..	..	..	..
Other	..	..	..	..	..	..
Main activity	287	287	287	287	287	287
Combustible fuels	1	1	*1	*1	*1	*1
Hydro	286	286	286	286	286	286
Nuclear	..	..	..	..	..	..
Other	..	..	..	..	..	..
Combustible fuel input	**Terajoules**					
Gas-diesel oil	*2012	*2012	*2012	*2012	*2012	*2012
Total input	*2012	*2012	*2012	*2012	*2012	*2012
Total production	*652	*652	*652	*652	*652	*652
Estimated efficiency (% of production to input)	32	32	32	32	32	32

126

Statistics on electricity

Malaysia

Item	2009	2010	2011	2012	2013	2014
Production, trade and consumption	Million kilowatt-hours					
Total main activity and autoproducer	107116	116808	124894	134421	138331	147461
Combustible fuels	100491	110556	117269	125319	127604	133846
Hydro	6625	6251	7624	9054	10586	13388
Nuclear	..	..	..	..	..	..
Other	..	1	1	48	141	227
Main activity	105203	114920	119761	128267	133311	142788
Combustible fuels	98578	108668	112136	119165	122584	129173
Hydro	6625	6251	7624	9054	10586	13388
Nuclear	..	..	..	..	..	..
Other	..	1	1	48	141	227
Own use in electricity, CHP and heat plants	1914	1892	12779	11543	11366	15585
Net production	105202	114916	112115	122878	126965	131876
Imports	0	0	370	100	221	23
Exports	92	148	10	15	23	12
Losses	3992	8200	0	0	0	0
Consumption	*96360	*110853	107341	116353	123076	128333
Energy industries own use	..	..	..	..	..	..
By industry and construction	43244	52715	47017	52414	55886	58948
By transport	*140	*209	214	241	241	256
By households and other cons.	*52976	*57929	60110	63698	66949	69129
Net installed capacity	Thousand kilowatts					
Total main activity and autoproducer	*25398	*25399	28750	29198	31817	29974
Combustible fuels	*23278	*23278	25734	25826	27725	25040
Hydro	*2120	*2120	3015	3317	3931	4773
Nuclear	..	..	..	..	..	..
Other	..	*1	*1	*55	*161	161
Main activity	*23993	*23994	25752	26024	28780	27623
Combustible fuels	*21873	*21873	22736	22652	24688	22695
Hydro	*2120	*2120	3015	3317	3931	4768
Nuclear	..	..	..	..	..	..
Other	..	*1	*1	*55	*161	160
Combustible fuel input	Terajoules					
Hard Coal	443683	542238	544825	591912	566336	571404
Gas-diesel oil	20167	26101	49966	35346	27348	27391
Fuel oil	*8363	*5090	44925	22382	15958	10948
Natural gas	618560	754903	531627	555449	695828	650110
Vegetal waste	*16762	*14400	*15048	*11405	*15955	9346
Total input	1107535	1342733	1186391	1216493	1321425	1269199
Total production	361767	398003	422168	451148	459374	481846
Estimated efficiency (% of production to input)	33	30	36	37	35	38

Maldives

Item	2009	2010	2011	2012	2013	2014
Production, trade and consumption	**Million kilowatt-hours**					
Total main activity and autoproducer	298	257	275	299	305	329
Combustible fuels	298	256	274	297	303	327
Hydro	..	..	..	..	..	..
Nuclear	..	..	..	..	..	..
Other	..	1	1	2	*2	*2
Main activity	298	257	275	299	305	329
Combustible fuels	298	256	274	297	303	327
Hydro	..	..	..	..	..	..
Nuclear	..	..	..	..	..	..
Other	..	1	1	2	*2	*2
Own use in electricity, CHP and heat plants	10	10	10	12	15	18
Net production	288	247	265	287	290	311
Imports	..	..	..	..	..	..
Exports	..	..	..	..	..	..
Losses	*19	*15	*15	*16	*16	*16
Consumption	*270	*232	*250	*271	*273	*295
Energy industries own use	..	..	..	..	..	..
By industry and construction	..	..	..	..	..	..
By transport	..	..	..	..	..	..
By households and other cons.	*270	*232	*250	*271	*273	*295
Net installed capacity	**Thousand kilowatts**					
Total main activity and autoproducer	75	78	81	84	85	271
Combustible fuels	75	77	78	81	81	266
Hydro	..	..	..	..	..	..
Nuclear	..	..	..	..	..	..
Other	..	*1	*3	*3	*5	5
Main activity	75	78	81	84	85	271
Combustible fuels	75	77	78	81	81	266
Hydro	..	..	..	..	..	..
Nuclear	..	..	..	..	..	..
Other	..	*1	*3	*3	*5	5
Combustible fuel input	**Terajoules**					
Gas-diesel oil	3225	2765	*2993	*3216	*3225	*4042
Total input	3225	2765	*2993	*3216	*3225	*4042
Total production	1074	921	985	1070	1090	1177
Estimated efficiency (% of production to input)	33	33	33	33	34	29

Statistics on electricity

Mali

Item	2009	2010	2011	2012	2013	2014
Production, trade and consumption	**Million kilowatt-hours**					
Total main activity and autoproducer	1175	1292	1381	1363	1510	*1679
Combustible fuels	323	597	654	579	532	*959
Hydro	851	695	727	783	978	*720
Nuclear	..	..	..	..	..	..
Other	..	..	..	..	..	..
Main activity	1123	1240	1329	1311	1458	*1627
Combustible fuels	271	545	602	527	480	*907
Hydro	851	695	727	783	978	*720
Nuclear	..	..	..	..	..	..
Other	..	..	..	..	..	..
Own use in electricity, CHP and heat plants	0	0	0	0	0	0
Net production	1175	1292	1381	1363	1510	*1679
Imports	..	..	..	..	..	..
Exports	348	368	*368	*368	*300	*300
Losses	*123	*132	*133	*135	131	*140
Consumption	1104	1191	*1201	*1215	1281	*1340
Energy industries own use	..	..	..	..	..	..
By industry and construction	625	639	*640	*640	641	*670
By transport	..	..	..	..	..	..
By households and other cons.	*479	*552	*561	*575	640	*670
Net installed capacity	**Thousand kilowatts**					
Total main activity and autoproducer	540	579	655	689	811	*890
Combustible fuels	193	225	291	323	363	*400
Hydro	348	354	365	367	448	*490
Nuclear	..	..	..	..	..	..
Other	..	..	..	..	..	..
Main activity	443	482	558	592	714	*780
Combustible fuels	96	128	194	226	266	*290
Hydro	348	354	365	367	448	*490
Nuclear	..	..	..	..	..	..
Other	..	..	..	..	..	..
Combustible fuel input	**Terajoules**					
Gas-diesel oil	*3956	*6063	*7095	*6278	*6536	*11954
Fuel oil	*808	*808	*808	*808	*808	*808
Total input	*4764	*6871	*7903	*7086	*7344	*12762
Total production	1164	2149	2356	2085	1914	*3454
Estimated efficiency (% of production to input)	24	31	30	29	26	27

Statistics on electricity

Malta

Item	2009	2010	2011	2012	2013	2014
Production, trade and consumption	**Million kilowatt-hours**					
Total main activity and autoproducer	2168	2116	2187	2307	2251	2245
Combustible fuels	2168	2115	2174	2277	2222	2177
Hydro	..	..	..	..	..	..
Nuclear	..	..	..	..	..	..
Other	0	1	13	30	29	68
Main activity	2168	2113	2177	2281	2216	2170
Combustible fuels	2168	2113	2169	2268	2216	2170
Hydro	..	..	..	..	..	..
Nuclear	..	..	..	..	..	..
Other	0	0	8	13	0	..
Own use in electricity, CHP and heat plants	122	123	134	135	114	108
Net production	2046	1993	2053	2172	2137	2137
Imports	..	..	..	..	..	..
Exports	..	..	..	..	..	..
Losses	339	138	153	193	157	105
Consumption	1707	1855	1900	1979	1980	2032
Energy industries own use	..	..	..	..	..	..
By industry and construction	509	407	415	417	410	413
By transport	..	..	..	..	..	..
By households and other cons.	1198	1448	1485	1562	1570	1619
Net installed capacity	**Thousand kilowatts**					
Total main activity and autoproducer	572	575	581	644	671	698
Combustible fuels	572	573	574	623	623	623
Hydro	..	..	..	..	..	..
Nuclear	..	..	..	..	..	..
Other	..	2	7	21	48	75
Main activity	571	572	576	636	648	675
Combustible fuels	571	571	571	620	620	620
Hydro	..	..	..	..	..	..
Nuclear	..	..	..	..	..	..
Other	..	1	5	16	28	55
Combustible fuel input	**Terajoules**					
Biogas	..	..	32	57	34	36
Gas-diesel oil	3311	3569	3139	3139	2494	2322
Fuel oil	21129	20846	21856	23311	18988	18665
Total input	24440	24415	25027	26507	21516	21023
Total production	7805	7614	7826	8197	7999	7837
Estimated efficiency (% of production to input)	32	31	31	31	37	37

Statistics on electricity

Marshall Islands

Item	2009	2010	2011	2012	2013	2014
Production, trade and consumption	Million kilowatt-hours					
Total main activity and autoproducer	*93	*92	*79	78	84	*85
Combustible fuels	*93	*92	*79	78	83	*85
Hydro	..	..	..	..	..	..
Nuclear	..	..	..	..	..	..
Other	..	..	..	0	0	0
Main activity	*93	*92	*79	78	84	*85
Combustible fuels	*93	*92	*79	78	83	*85
Hydro	..	..	..	..	..	..
Nuclear	..	..	..	..	..	..
Other	..	..	..	0	0	0
Own use in electricity, CHP and heat plants	0	0	0	0	0	0
Net production	*93	*92	*79	78	84	*85
Imports	..	..	..	..	..	..
Exports	..	..	..	..	..	..
Losses	*11	*11	*10	10	*11	*11
Consumption	*82	*81	*69	68	*73	*74
Energy industries own use	..	..	..	..	..	..
By industry and construction	..	..	..	..	..	..
By transport	..	..	..	..	..	..
By households and other cons.	*82	*81	*69	68	*73	*74
Net installed capacity	Thousand kilowatts					
Total main activity and autoproducer	29	29	29	33	33	34
Combustible fuels	29	29	29	33	33	33
Hydro	..	..	..	..	..	..
Nuclear	..	..	..	..	..	..
Other	..	..	..	0	0	1
Main activity	29	29	29	33	33	34
Combustible fuels	29	29	29	33	33	33
Hydro	..	..	..	..	..	..
Nuclear	..	..	..	..	..	..
Other	..	..	..	0	0	1
Combustible fuel input	Terajoules					
Gas-diesel oil	*869	*847	*727	714	787	*796
Total input	*869	*847	*727	714	787	*796
Total production	*335	*331	*284	279	300	*306
Estimated efficiency (% of production to input)	39	39	39	39	38	38

Martinique

Item	2009	2010	2011	2012	2013	2014
Production, trade and consumption	**Million kilowatt-hours**					
Total main activity and autoproducer	*1732	*1808	*1764	*1780	*1764	*1764
Combustible fuels	*1685	*1764	*1720	*1736	*1719	*1719
Hydro	..	..	..	..	..	..
Nuclear	..	..	..	..	..	..
Other	*47	*45	*45	*45	*46	*46
Main activity	*1702	*1778	*1734	*1750	*1734	*1734
Combustible fuels	*1655	*1734	*1690	*1706	*1689	*1689
Hydro	..	..	..	..	..	..
Nuclear	..	..	..	..	..	..
Other	*47	*45	*45	*45	*46	*46
Own use in electricity, CHP and heat plants	*79	*191	*188	*189	*187	*202
Net production	*1653	*1617	*1576	*1591	*1577	*1562
Imports	..	..	..	..	..	..
Exports	..	..	..	..	..	..
Losses	*8	*8	*8	*8	*8	*8
Consumption	*1550	*1582	*1547	*1558	*1544	*1544
Energy industries own use	..	..	..	..	..	..
By industry and construction	*30	*30	*30	*30	*30	*30
By transport	..	..	..	..	..	..
By households and other cons.	*1520	*1552	*1517	*1528	*1514	*1514
Net installed capacity	**Thousand kilowatts**					
Total main activity and autoproducer	*415	*415	*415	*461	*463	492
Combustible fuels	*400	*400	*400	*400	*400	428
Hydro	..	..	..	..	..	..
Nuclear	..	..	..	..	..	..
Other	15	*15	*15	61	63	64
Main activity	*411	*411	*411	*457	*459	488
Combustible fuels	*396	*396	*396	*396	*396	424
Hydro	..	..	..	..	..	..
Nuclear	..	..	..	..	..	..
Other	15	*15	*15	61	63	64
Combustible fuel input	**Terajoules**					
Gas-diesel oil	*1118	*1075	*1075	*1032	*1032	*1118
Fuel oil	*12223	*12928	*13705	*13946	*14005	*14579
Municipal waste	*432	*432	*432	*432	*432	*432
Total input	*13773	*14435	*15212	*15410	*15469	*16129
Total production	*6068	*6350	*6191	*6249	*6188	*6188
Estimated efficiency (% of production to input)	44	44	41	41	40	38

132

Statistics on electricity

Mauritania

Item	2009	2010	2011	2012	2013	2014
Production, trade and consumption	**Million kilowatt-hours**					
Total main activity and autoproducer	658	727	744	824	790	910
Combustible fuels	658	727	744	824	790	910
Hydro	..	..	..	..	..	..
Nuclear	..	..	..	..	..	..
Other	..	..	..	..	..	..
Main activity	381	426	436	501	470	530
Combustible fuels	381	426	436	501	470	530
Hydro	..	..	..	..	..	..
Nuclear	..	..	..	..	..	..
Other	..	..	..	..	..	..
Own use in electricity, CHP and heat plants	10	11	14	18	35	155
Net production	648	716	731	806	755	755
Imports	109	115	*115	*115	*115	135
Exports	..	..	..	..	..	..
Losses	130	148	139	141	164	*164
Consumption	623	669	698	771	706	862
Energy industries own use	..	..	..	..	..	..
By industry and construction	276	301	309	323	212	*212
By transport	..	..	..	..	..	..
By households and other cons.	346	368	390	447	494	650
Net installed capacity	**Thousand kilowatts**					
Total main activity and autoproducer	166	196	*196	*196	*263	*315
Combustible fuels	166	196	*196	*196	*263	*315
Hydro	..	..	..	..	..	..
Nuclear	..	..	..	..	..	..
Other	..	..	..	..	..	..
Main activity	96	126	*126	*126	*193	*236
Combustible fuels	96	126	*126	*126	*193	*236
Hydro	..	..	..	..	..	..
Nuclear	..	..	..	..	..	..
Other	..	..	..	..	..	..
Combustible fuel input	**Terajoules**					
Gas-diesel oil	3698	3410	*4300	*5010	5010	*5010
Fuel oil	2351	2727	*2828	*3232	*3232	*3232
Total input	6049	6137	*7128	*8242	8242	*8242
Total production	2367	2618	2679	2967	2844	3276
Estimated efficiency (% of production to input)	39	43	38	36	35	40

Statistics on electricity

Mauritius

Item	2009	2010	2011	2012	2013	2014
Production, trade and consumption	**Million kilowatt-hours**					
Total main activity and autoproducer	2577	2689	2739	2798	2882	2936
Combustible fuels	2454	2585	2679	2719	2781	2817
Hydro	122	101	57	74	95	91
Nuclear	..	..	..	..	..	..
Other	2	3	3	5	6	28
Main activity	1077	1099	1130	1146	1176	1175
Combustible fuels	954	995	1070	1068	1078	1081
Hydro	122	101	57	74	95	91
Nuclear	..	..	..	..	..	..
Other	2	3	3	4	4	3
Own use in electricity, CHP and heat plants	39	40	44	43	42	43
Net production	2538	2649	2695	2754	2840	2893
Imports	..	..	..	..	..	..
Exports	..	..	..	..	..	..
Losses	198	194	194	191	184	182
Consumption	2340	2456	2491	2561	2657	2708
Energy industries own use	272	281	264	267	273	256
By industry and construction	*659	*686	*682	*687	*714	*715
By transport	..	..	..	..	..	..
By households and other cons.	1409	1489	1546	1607	1670	1737
Net installed capacity	**Thousand kilowatts**					
Total main activity and autoproducer	740	741	738	781	778	783
Combustible fuels	680	680	677	718	714	703
Hydro	59	60	60	60	61	61
Nuclear	..	..	..	..	..	..
Other	*1	*1	1	3	4	19
Main activity	443	444	450	506	501	490
Combustible fuels	383	383	389	445	439	428
Hydro	59	60	60	60	61	61
Nuclear	..	..	..	..	..	..
Other	*1	*1	1	1	1	1
Combustible fuel input	**Terajoules**					
Hard Coal	14901	16692	16017	16848	17731	18344
Biogas	..	..	..	..	240	..
Gas-diesel oil	129	86	86	86	43	43
Fuel oil	7716	7959	8686	8605	8726	8928
Other kerosene	219	263	175	131	44	44
Bagasse	8772	8803	8641	8324	8155	7954
Total input	31738	33803	33606	33995	34938	35313
Total production	8833	9307	9645	9788	10012	10142
Estimated efficiency (% of production to input)	28	28	29	29	29	29

Statistics on electricity

Mayotte

Item	2009	2010	2011	2012	2013	2014
Production, trade and consumption	Million kilowatt-hours					
Total main activity and autoproducer	239	258	262	271	285	299
Combustible fuels	237	251	248	256	268	283
Hydro	..	..	..	..	..	..
Nuclear	..	..	..	..	..	..
Other	1	6	14	15	17	17
Main activity	239	258	262	271	285	299
Combustible fuels	237	251	248	256	268	283
Hydro	..	..	..	..	..	..
Nuclear	..	..	..	..	..	..
Other	1	6	14	15	17	17
Own use in electricity, CHP and heat plants	13	12	12	14	15	16
Net production	226	246	250	257	270	283
Imports	..	..	..	..	..	..
Exports	..	..	..	..	..	..
Losses	*9	*10	*10	11	10	10
Consumption	217	236	240	246	261	274
Energy industries own use	..	..	..	..	..	..
By industry and construction	..	..	..	..	..	..
By transport	..	..	..	..	..	..
By households and other cons.	217	236	240	246	261	274
Net installed capacity	Thousand kilowatts					
Total main activity and autoproducer	80	86	90	91	91	*91
Combustible fuels	78	78	78	78	78	*78
Hydro	..	..	..	..	..	..
Nuclear	..	..	..	..	..	..
Other	2	8	12	*13	13	13
Main activity	80	86	90	91	91	*91
Combustible fuels	78	78	78	78	78	*78
Hydro	..	..	..	..	..	..
Nuclear	..	..	..	..	..	..
Other	2	8	12	*13	13	13
Combustible fuel input	Terajoules					
Gas-diesel oil	2172	2292	2232	*2322	*2438	*2559
Total input	2172	2292	2232	*2322	*2438	*2559
Total production	855	905	893	920	966	1018
Estimated efficiency (% of production to input)	39	39	40	40	40	40

Statistics on electricity

Mexico

Item	2009	2010	2011	2012	2013	2014
Production, trade and consumption	**Million kilowatt-hours**					
Total main activity and autoproducer	267754	275537	302785	307268	297326	301496
Combustible fuels	223172	224639	248253	257041	247163	240279
Hydro	26718	37131	36247	31883	28002	38893
Nuclear	10501	5879	10089	8770	11800	9677
Other	7363	7888	8196	9574	10361	12647
Main activity	235106	242536	266359	262241	258821	258258
Combustible fuels	191171	193135	213610	214591	211680	202346
Hydro	26445	36738	35796	31317	27444	38145
Nuclear	10501	5879	10089	8770	11800	9677
Other	6989	6784	6864	7563	7897	8090
Own use in electricity, CHP and heat plants	11008	11201	22690	22863	10236	11446
Net production	256746	264336	280095	284405	287090	290050
Imports	346	397	596	2177	1210	2124
Exports	1249	1348	1293	1116	1240	2653
Losses	42452	44252	45602	44050	42520	41322
Consumption	211272	221091	238955	253065	246503	256779
Energy industries own use	5411	5397	5138	5355	5033	4559
By industry and construction	114201	122713	134551	146746	137391	142326
By transport	1116	1191	1122	1130	1130	1130
By households and other cons.	90544	91790	98144	99834	102949	108764
Net installed capacity	**Thousand kilowatts**					
Total main activity and autoproducer	59574	61389	62046	64077	64092	66238
Combustible fuels	45320	46914	47586	48364	48047	48893
Hydro	11474	11597	11571	11626	11633	12464
Nuclear	1365	1365	1365	1400	1400	1400
Other	1415	1513	1524	2687	3012	3481
Main activity	51686	52591	52465	52336	53497	54379
Combustible fuels	37888	38673	38673	38015	39161	39294
Hydro	11383	11503	11453	11498	11509	12269
Nuclear	1365	1365	1365	1400	1400	1400
Other	1050	1050	974	1423	1427	1416
Combustible fuel input	**Terajoules**					
Hard Coal	147362	127036	161571	139230	125044	150551
Brown Coal	159396	195484	188607	205642	197886	200358
Biogas	823	1298	1471	1822	1970	1939
Gas-diesel oil	22876	18920	20640	25456	35475	21371
Fuel oil	412040	387800	436199	454581	432522	300010
Liquefied petroleum gas	378	662	3122	1372	568	5061
Petroleum coke	31753	39163	37928	39293	40073	39910
Natural gas	1364692	1363614	1453310	1523582	1423071	1484833
Blast furnace gas	5045	6712	5071	3497	4199	4620
Bagasse	30448	41035	31790	41468	53758	71125
Others	1266	1377	1605	9427	2454	1347
Total input	2176079	2183100	2341313	2445369	2317019	2281126
Total production	803419	808700	893711	925348	889787	865004
Estimated efficiency (% of production to input)	37	37	38	38	38	38

136

Statistics on electricity

Micronesia (Fed. States of)

Item	2009	2010	2011	2012	2013	2014
Production, trade and consumption	Million kilowatt-hours					
Total main activity and autoproducer	70	70	66	*67	*68	70
Combustible fuels	69	68	64	*65	*66	68
Hydro	..	..	..	..	..	..
Nuclear	..	..	..	..	..	..
Other	*1	*2	*2	*2	*2	*2
Main activity	70	70	66	*67	*68	70
Combustible fuels	69	68	64	*65	*66	68
Hydro	..	..	..	..	..	..
Nuclear	..	..	..	..	..	..
Other	*1	*2	*2	*2	*2	*2
Own use in electricity, CHP and heat plants	0	0	0	0	0	0
Net production	70	70	66	*67	*68	70
Imports	..	..	..	..	..	..
Exports	..	..	..	..	..	..
Losses	14	14	13	*13	*13	*15
Consumption	*56	*56	*53	*54	*55	*55
Energy industries own use	..	..	..	..	..	..
By industry and construction	*4	*4	*4	*4	*4	*4
By transport	..	..	..	..	..	..
By households and other cons.	*52	*52	*49	*50	*51	*51
Net installed capacity	Thousand kilowatts					
Total main activity and autoproducer	33	33	33	33	33	34
Combustible fuels	32	32	32	32	32	32
Hydro	..	..	..	..	..	2
Nuclear	..	..	..	..	..	..
Other	*1	*2	*2	*2	1	1
Main activity	33	33	33	33	33	34
Combustible fuels	32	32	32	32	32	32
Hydro	..	..	..	..	..	2
Nuclear	..	..	..	..	..	..
Other	*1	*2	*2	*2	1	1
Combustible fuel input	Terajoules					
Gas-diesel oil	730	731	673	*688	*688	*688
Total input	730	731	673	*688	*688	*688
Total production	249	246	230	*235	*238	243
Estimated efficiency (% of production to input)	34	34	34	34	35	35

Statistics on electricity

Mongolia

Item	2009	2010	2011	2012	2013	2014
Production, trade and consumption	**Million kilowatt-hours**					
Total main activity and autoproducer	5182	6215	4812	5182	6215	6725
Combustible fuels	5182	6215	4812	5182	6215	6725
Hydro	..	..	..	..	..	..
Nuclear	..	..	..	..	..	..
Other	..	..	..	..	..	..
Main activity	4816	5020	4536	4816	5020	5376
Combustible fuels	4816	5020	4536	4816	5020	5376
Hydro	..	..	..	..	..	..
Nuclear	..	..	..	..	..	..
Other	..	..	..	..	..	..
Own use in electricity, CHP and heat plants	712	725	691	712	725	772
Net production	4469	5490	4121	4469	5490	5953
Imports	366	1195	276	366	1196	1349
Exports	21	18	24	21	18	33
Losses	675	740	644	675	740	793
Consumption	3773	4732	3453	3773	4732	5159
Energy industries own use	..	..	..	..	..	..
By industry and construction	2339	2931	2141	2339	2931	3172
By transport	157	197	144	157	197	211
By households and other cons.	1277	1605	1169	1277	1605	1776
Net installed capacity	**Thousand kilowatts**					
Total main activity and autoproducer	832	*922	*1340	*1340	*1340	*1370
Combustible fuels	832	*922	*1340	*1340	*1340	*1370
Hydro	..	..	..	..	..	..
Nuclear	..	..	..	..	..	..
Other	..	..	..	..	..	..
Main activity	772	*772	*1190	*1190	*1190	*1190
Combustible fuels	772	*772	*1190	*1190	*1190	*1190
Hydro	..	..	..	..	..	..
Nuclear	..	..	..	..	..	..
Other	..	..	..	..	..	..
Combustible fuel input	**Terajoules**					
Hard Coal	16619	14362	26579	30620	35173	38740
Brown Coal	63006	72425	64469	68053	74013	76185
Gas-diesel oil	2021	2408	3096	2881	3741	3311
Fuel oil	202	202	283	283	162	125
Total input	81848	89398	94427	101836	113088	118361
Total production	18654	22374	17323	18654	22374	24210
Estimated efficiency (% of production to input)	23	25	18	18	20	20

Statistics on electricity

Montenegro

Item	2009	2010	2011	2012	2013	2014
Production, trade and consumption	Million kilowatt-hours					
Total main activity and autoproducer	2841	4168	2809	2844	3945	3040
Combustible fuels	768	1413	1600	1367	1444	1322
Hydro	2073	2755	1209	1477	2501	1716
Nuclear	..	..	..	..	..	..
Other	..	..	..	..	..	2
Main activity	2841	4168	2809	2844	3945	3038
Combustible fuels	768	1413	1600	1367	1444	1322
Hydro	2073	2755	1209	1477	2501	1716
Nuclear	..	..	..	..	..	..
Other	..	..	..	..	..	..
Own use in electricity, CHP and heat plants	81	146	153	141	136	135
Net production	2760	4022	2656	2703	3809	2905
Imports	1151	732	1383	964	388	903
Exports	172	483	431	228	647	642
Losses	717	667	651	695	622	557
Consumption	3017	3211	3414	3220	2706	2610
Energy industries own use	..	..	..	..	..	..
By industry and construction	1694	1853	2162	1855	1326	760
By transport	18	22	20	24	34	32
By households and other cons.	1305	1336	1232	1341	1346	1818
Net installed capacity	Thousand kilowatts					
Total main activity and autoproducer	868	868	868	868	868	870
Combustible fuels	210	210	210	210	210	210
Hydro	658	658	658	658	658	658
Nuclear	..	..	..	..	..	..
Other	..	..	..	..	..	2
Main activity	868	868	868	868	868	868
Combustible fuels	210	210	210	210	210	210
Hydro	658	658	658	658	658	658
Nuclear	..	..	..	..	..	..
Other	..	..	..	..	..	..
Combustible fuel input	Terajoules					
Brown Coal	8076	17039	17499	15611	15178	14708
Total input	8076	17039	17499	15611	15178	14708
Total production	2765	5087	5760	4921	5198	4759
Estimated efficiency (% of production to input)	34	30	33	32	34	32

Statistics on electricity

Montserrat

Item	2009	2010	2011	2012	2013	2014
Production, trade and consumption	Million kilowatt-hours					
Total main activity and autoproducer	*23	*23	*23	*23	*23	24
Combustible fuels	*23	*23	*23	*23	*23	24
Hydro	..	..	..	..	..	..
Nuclear	..	..	..	..	..	..
Other	..	..	..	..	..	..
Main activity	*23	*23	*23	*23	*23	24
Combustible fuels	*23	*23	*23	*23	*23	24
Hydro	..	..	..	..	..	..
Nuclear	..	..	..	..	..	..
Other	..	..	..	..	..	..
Own use in electricity, CHP and heat plants	*2	*2	*2	*2	*2	2
Net production	*21	*21	*21	*21	*21	22
Imports	..	..	..	..	..	..
Exports	..	..	..	..	..	..
Losses	*2	*2	*2	*2	*2	*3
Consumption	*19	*19	19	19	20	20
Energy industries own use	..	..	..	..	..	..
By industry and construction	*9	*9	*9	*9	*9	*9
By transport	..	..	..	..	..	..
By households and other cons.	*10	*10	10	10	11	11
Net installed capacity	Thousand kilowatts					
Total main activity and autoproducer	*5	*5	*5	*5	*5	*5
Combustible fuels	*5	*5	*5	*5	*5	*5
Hydro	..	..	..	..	..	..
Nuclear	..	..	..	..	..	..
Other	..	..	..	..	..	..
Main activity	*5	*5	*5	*5	*5	*5
Combustible fuels	*5	*5	*5	*5	*5	*5
Hydro	..	..	..	..	..	..
Nuclear	..	..	..	..	..	..
Other	..	..	..	..	..	..
Combustible fuel input	Terajoules					
Gas-diesel oil	*237	*237	*237	*237	*237	*237
Total input	*237	*237	*237	*237	*237	*237
Total production	*83	*83	*83	*83	*83	86
Estimated efficiency (% of production to input)	35	35	35	35	35	37

Statistics on electricity

Morocco

Item	2009	2010	2011	2012	2013	2014
Production, trade and consumption	Million kilowatt-hours					
Total main activity and autoproducer	20935	22853	24364	26495	27929	29142
Combustible fuels	17592	18563	21533	23951	23663	25185
Hydro	2952	3631	2139	1816	2785	2033
Nuclear	..	..	..	..	..	..
Other	391	659	692	728	1481	1924
Main activity	20809	22700	24147	26356	26749	27945
Combustible fuels	17466	18410	21316	23812	22483	23988
Hydro	2952	3631	2139	1816	2785	2033
Nuclear	..	..	..	..	..	..
Other	391	659	692	728	1481	1924
Own use in electricity, CHP and heat plants	37	43	40	40	489	547
Net production	20898	22810	24324	26455	27440	28595
Imports	4623	3939	4607	4841	5551	6138
Exports	..	..	..	..	151	128
Losses	2632	2825	3082	3496	3611	4227
Consumption	22889	23924	25849	27800	29229	30378
Energy industries own use	505	219	179	241	1383	1572
By industry and construction	8505	9008	9741	*10700	10204	10505
By transport	230	280	303	301	322	332
By households and other cons.	13649	14417	15626	16558	17320	17969
Net installed capacity	Thousand kilowatts					
Total main activity and autoproducer	6466	6679	6713	7013	*7013	*7263
Combustible fuels	4496	4688	4693	4993	*4993	*4993
Hydro	1749	1770	1770	1770	*1770	*1770
Nuclear	..	..	..	..	..	..
Other	221	221	250	250	*250	*500
Main activity	6135	6348	6377	6677	*6677	*6927
Combustible fuels	4165	4357	4357	4657	*4657	*4657
Hydro	1749	1770	1770	1770	*1770	*1770
Nuclear	..	..	..	..	..	..
Other	221	221	250	250	*250	*500
Combustible fuel input	Terajoules					
Hard Coal	112632	119264	125813	126200	123547	168285
Gas-diesel oil	602	1419	1720	989	903	645
Fuel oil	39188	53207	64882	66066	51106	28765
Petroleum coke	1300	0	0	0	..	..
Natural gas	22818	24403	31203	42409	44289	42883
Total input	176540	198293	223618	235664	219845	240578
Total production	63331	66825	77520	86225	85187	90666
Estimated efficiency (% of production to input)	36	34	35	37	39	38

Statistics on electricity

Mozambique

Item	2009	2010	2011	2012	2013	2014
Production, trade and consumption	**Million kilowatt-hours**					
Total main activity and autoproducer	16963	16666	16830	15166	15645	17739
Combustible fuels	13	19	20	21	840	1379
Hydro	16950	16647	16810	15145	14805	16359
Nuclear	..	..	..	..	..	..
Other	..	..	..	..	..	1
Main activity	16963	16309	16482	14903	15382	17271
Combustible fuels	13	19	20	21	840	1379
Hydro	16950	16290	16462	14882	14542	15892
Nuclear	..	..	..	..	..	..
Other	..	..	..	..	..	..
Own use in electricity, CHP and heat plants	143	144	149	210	220	238
Net production	16820	16522	16681	14956	15425	17501
Imports	8340	8533	8570	8304	8339	7656
Exports	12700	12075	11954	9791	9058	10199
Losses	2443	2457	2456	2243	2647	2612
Consumption	9569	9831	10141	10939	11605	12342
Energy industries own use	..	..	..	..	..	..
By industry and construction	8466	8574	8681	9448	9295	10651
By transport	..	..	..	..	..	..
By households and other cons.	1103	1257	1460	1491	2310	1691
Net installed capacity	**Thousand kilowatts**					
Total main activity and autoproducer	2428	2478	2479	2486	2646	2682
Combustible fuels	249	249	249	249	359	359
Hydro	2179	2229	2230	2237	2287	2322
Nuclear	..	..	..	..	..	..
Other	..	..	..	..	..	1
Main activity	2428	2428	2429	2436	2596	2596
Combustible fuels	249	249	249	249	359	359
Hydro	2179	2179	2180	2187	2237	2237
Nuclear	..	..	..	..	..	..
Other	..	..	..	..	..	..
Combustible fuel input	**Terajoules**					
Gas-diesel oil	..	..	..	..	*6880	15980
Natural gas	170	215	304	357	3026	16137
Total input	170	215	304	357	*9906	32117
Total production	47	68	72	76	3024	4964
Estimated efficiency (% of production to input)	28	32	24	21	31	15

Statistics on electricity

Myanmar

Item	2009	2010	2011	2012	2013	2014
Production, trade and consumption	Million kilowatt-hours					
Total main activity and autoproducer	6964	8625	10425	10965	12247	14156
Combustible fuels	1708	2436	2907	3199	3424	5327
Hydro	5256	6189	7518	7766	8823	8829
Nuclear	..	..	..	..	..	..
Other	..	..	..	..	..	..
Main activity	6964	8625	10425	10923	11302	11197
Combustible fuels	1708	2436	2907	3199	2929	3046
Hydro	5256	6189	7518	7725	8373	8151
Nuclear	..	..	..	..	..	..
Other	..	..	..	..	..	..
Own use in electricity, CHP and heat plants	165	155	160	186	182	152
Net production	6799	8470	10265	10779	12065	14005
Imports	..	..	..	..	..	..
Exports	..	..	..	..	..	..
Losses	1856	2158	2548	2524	2452	2750
Consumption	4943	6312	7717	8255	9613	11255
Energy industries own use	..	..	..	..	..	..
By industry and construction	2928	3659	4336	4600	5849	7142
By transport	..	..	..	..	..	..
By households and other cons.	2015	2653	3381	3655	3764	4113
Net installed capacity	Thousand kilowatts					
Total main activity and autoproducer	2544	3413	3588	3726	4146	4805
Combustible fuels	890	891	895	913	1141	1620
Hydro	1654	2522	2693	2813	3005	3185
Nuclear	..	..	..	..	..	..
Other	..	..	..	..	..	..
Main activity	2544	3413	3588	3606	3753	4177
Combustible fuels	890	891	895	913	920	1164
Hydro	1654	2522	2693	2693	2833	3013
Nuclear	..	..	..	..	..	..
Other	..	..	..	..	..	..
Combustible fuel input	Terajoules					
Hard Coal	4745	6745	6745	7048	4699	2628
Gas-diesel oil	1247	2150	1634	3440	2580	4300
Natural gas	17393	25029	22922	30948	40331	71842
Total input	23385	33924	31301	41436	47610	78770
Total production	6148	8770	10465	11515	12326	19179
Estimated efficiency (% of production to input)	26	26	33	28	26	24

Statistics on electricity

Namibia

Item	2009	2010	2011	2012	2013	2014
Production, trade and consumption	**Million kilowatt-hours**					
Total main activity and autoproducer	1742	1488	1607	1538	1712	1498
Combustible fuels	313	224	9	67	72	13
Hydro	1429	1264	1598	1471	1640	1485
Nuclear	..	..	..	..	..	..
Other	..	..	..	..	..	..
Main activity	1742	1488	1607	1538	1712	1498
Combustible fuels	313	224	9	67	72	13
Hydro	1429	1264	1598	1471	1640	1485
Nuclear	..	..	..	..	..	..
Other	..	..	..	..	..	..
Own use in electricity, CHP and heat plants	0	0	0	0	0	0
Net production	1742	1488	1607	1538	1712	1498
Imports	2202	2462	2389	2226	2916	2886
Exports	144	207	79	88	89	84
Losses	255	329	360	426	369	543
Consumption	3290	3354	3467	3635	3772	3747
Energy industries own use	..	..	..	..	..	..
By industry and construction	761	803	817	795	786	716
By transport	..	..	..	..	..	..
By households and other cons.	2529	2551	2650	2840	2986	3031
Net installed capacity	**Thousand kilowatts**					
Total main activity and autoproducer	*467	*467	393	508	487	487
Combustible fuels	*92	*92	144	167	155	155
Hydro	*375	*375	249	341	332	332
Nuclear	..	..	..	..	..	..
Other	..	..	..	..	..	..
Main activity	*467	*467	393	508	487	487
Combustible fuels	*92	*92	144	167	155	155
Hydro	*375	*375	249	341	332	332
Nuclear	..	..	..	..	..	..
Other	..	..	..	..	..	..
Combustible fuel input	**Terajoules**					
Hard Coal	4380	3086	136	726	590	0
Gas-diesel oil	86	86	*86	*86	*86	172
Total input	4466	3172	222	812	676	172
Total production	1127	806	32	241	259	47
Estimated efficiency (% of production to input)	25	25	15	30	38	27

Statistics on electricity

Nauru

Item	2009	2010	2011	2012	2013	2014
Production, trade and consumption	Million kilowatt-hours					
Total main activity and autoproducer	21	23	23	24	*24	*25
Combustible fuels	21	23	23	24	*24	*25
Hydro	..	..	..	..	..	..
Nuclear	..	..	..	..	..	..
Other	0	0	0	0	0	0
Main activity	21	23	23	24	*24	*25
Combustible fuels	21	23	23	24	*24	*25
Hydro	..	..	..	..	..	..
Nuclear	..	..	..	..	..	..
Other	0	0	0	0	0	0
Own use in electricity, CHP and heat plants	0	0	0	0	0	0
Net production	21	23	23	24	*24	*25
Imports	..	..	..	..	..	..
Exports	..	..	..	..	..	..
Losses	*2	*2	*2	*2	*2	*2
Consumption	*19	*20	21	*21	*22	*23
Energy industries own use	..	..	..	..	..	..
By industry and construction	..	..	..	..	..	..
By transport	..	..	..	..	..	..
By households and other cons.	*19	*20	21	*21	*22	*23
Net installed capacity	Thousand kilowatts					
Total main activity and autoproducer	*6	*6	5	5	*5	*6
Combustible fuels	*6	*6	5	5	*5	*6
Hydro	..	..	..	..	..	..
Nuclear	..	..	..	..	..	..
Other	0	0	0	0	0	0
Main activity	*6	*6	5	5	*5	*6
Combustible fuels	*6	*6	5	5	*5	*6
Hydro	..	..	..	..	..	..
Nuclear	..	..	..	..	..	..
Other	0	0	0	0	0	0
Combustible fuel input	Terajoules					
Gas-diesel oil	235	269	276	283	*292	*305
Total input	235	269	276	283	*292	*305
Total production	76	81	83	85	*87	*90
Estimated efficiency (% of production to input)	32	30	30	30	30	29

Statistics on electricity

Nepal

Item	2009	2010	2011	2012	2013	2014
Production, trade and consumption	**Million kilowatt-hours**					
Total main activity and autoproducer	3115	3208	3492	3543	3130	3508
Combustible fuels	13	3	2	10	10	10
Hydro	3102	3205	3490	3533	3120	3498
Nuclear	..	..	..	..	..	..
Other	..	..	..	..	..	..
Main activity	3073	3164	3433	3459	2993	3368
Combustible fuels	13	3	2	10	10	10
Hydro	3060	3161	3431	3449	2983	3358
Nuclear	..	..	..	..	..	..
Other	..	..	..	..	..	..
Own use in electricity, CHP and heat plants	37	30	32	24	40	15
Net production	3078	3178	3460	3519	3090	3493
Imports	639	694	746	793	1072	1319
Exports	75	31	4	4	4	3
Losses	1073	1101	1102	1066	745	*750
Consumption	2567	2740	3099	3240	3413	3491
Energy industries own use	..	..	..	..	..	..
By industry and construction	960	1002	1124	1168	1222	1252
By transport	5	6	7	8	6	6
By households and other cons.	1602	1732	1968	2064	2185	2233
Net installed capacity	**Thousand kilowatts**					
Total main activity and autoproducer	*630	*630	*762	*762	*772	*829
Combustible fuels	*57	*57	53	53	53	53
Hydro	*573	*573	*709	*709	*718	*776
Nuclear	..	..	..	..	..	..
Other	..	..	..	..	..	..
Main activity	*475	*475	*607	*607	*617	*654
Combustible fuels	*57	*57	53	53	53	53
Hydro	*418	*418	*554	*554	*563	*601
Nuclear	..	..	..	..	..	..
Other	..	..	..	..	..	..
Combustible fuel input	**Terajoules**					
Fuel oil	105	57	16	101	89	*89
Total input	105	57	16	101	89	*89
Total production	47	11	7	36	36	35
Estimated efficiency (% of production to input)	45	19	45	36	41	39

Netherlands

Item	2009	2010	2011	2012	2013	2014
Production, trade and consumption	Million kilowatt-hours					
Total main activity and autoproducer	113691	119270	113963	103298	101736	103418
Combustible fuels	104589	110992	104412	93925	92494	92488
Hydro	98	105	57	104	114	112
Nuclear	4248	3969	4141	3915	2891	4091
Other	4756	4204	5353	5354	6237	6727
Main activity	91783	95187	89215	78150	77128	81458
Combustible fuels	83667	87815	80830	70044	69496	72440
Hydro	98	105	57	104	114	112
Nuclear	4248	3969	4141	3915	2891	4091
Other	3770	3298	4187	4087	4627	4815
Own use in electricity, CHP and heat plants	4924	4444	4818	4651	4799	4644
Net production	108767	114826	109145	98647	96937	98774
Imports	15452	15583	20620	32156	33252	32855
Exports	10561	12808	11531	15046	15015	18128
Losses	5245	5633	5198	5191	5133	4934
Consumption	109048	112550	113229	108753	110025	107222
Energy industries own use	4054	4550	5081	4747	5652	5592
By industry and construction	36814	39460	39393	35180	34974	33289
By transport	1678	1756	1741	1763	1750	1716
By households and other cons.	66502	66784	67014	67063	67649	66625
Net installed capacity	Thousand kilowatts					
Total main activity and autoproducer	25993	26688	28053	29924	30539	31762
Combustible fuels	23083	23743	24970	26518	26520	27286
Hydro	37	37	37	37	37	37
Nuclear	510	510	510	510	485	485
Other	2363	2398	2536	2859	3497	3954
Main activity	20673	21342	22022	23714	24136	24969
Combustible fuels	18336	18993	19609	21203	21390	22092
Hydro	37	37	37	37	37	37
Nuclear	510	510	510	510	485	485
Other	1790	1802	1866	1964	2224	2355
Combustible fuel input	Terajoules					
Hard Coal	213427	202325	187705	217567	221265	257247
Biogas	7325	7851	7712	7108	6640	6169
Gas-diesel oil	516	473	430	688	1032	2924
Refinery gas	17870	15791	21434	17325	17375	20345
Other oil products	1246	1286	1809	1246	764	3055
Natural gas	589277	631129	559525	461008	450275	423496
Blast furnace gas	16675	23842	25281	25063	23532	24677
Fuelwood	*27400	*30348	*26937	*27330	*16651	*8767
Vegetal waste	*7850	*10440	*10947	*11448	*12466	*12888
Municipal waste	49219	48998	54776	67182	70182	70843
Others	2413	2523	2020	2237	2307	2005
Total input	933218	975005	898576	838202	822488	832416
Total production	376520	399571	375883	338130	332978	332957
Estimated efficiency (% of production to input)	40	41	42	40	40	40

Statistics on electricity

New Caledonia

Item	2009	2010	2011	2012	2013	2014
Production, trade and consumption	**Million kilowatt-hours**					
Total main activity and autoproducer	1945	2131	2267	2288	2506	2622
Combustible fuels	1505	1811	1843	1827	1994	2272
Hydro	397	265	375	399	455	288
Nuclear	..	..	..	..	..	..
Other	42	55	49	62	57	62
Main activity	1944	2131	2256	2264	2299	2380
Combustible fuels	1505	1811	1833	1804	1788	2031
Hydro	397	265	375	399	455	288
Nuclear						
Other	42	54	49	61	56	61
Own use in electricity, CHP and heat plants	5	23	9	29	1	63
Net production	1939	2108	2258	2259	2505	2559
Imports	..	..	..	..	..	..
Exports	..	..	..	..	..	..
Losses	36	60	106	43	107	105
Consumption	1903	2049	2152	2216	2398	2453
Energy industries own use	..	..	..	..	..	..
By industry and construction	1347	1469	1535	1580	1769	1826
By transport	..	..	..	..	..	..
By households and other cons.	556	*579	*617	*636	629	*627
Net installed capacity	**Thousand kilowatts**					
Total main activity and autoproducer	494	500	498	503	691	683
Combustible fuels	383	383	380	384	572	565
Hydro	78	78	78	78	78	78
Nuclear	..	..	..	..	..	..
Other	33	40	41	41	41	40
Main activity	493	500	494	495	630	622
Combustible fuels	383	383	377	377	512	505
Hydro	78	78	78	78	78	78
Nuclear	..	..	..	..	..	..
Other	32	39	40	41	41	40
Combustible fuel input	**Terajoules**					
Hard Coal	2609	*7783	*8037	*7867	8479	*9084
Gas-diesel oil	344	*344	*430	*559	2722	*2722
Fuel oil	13655	*12524	*12524	*12524	12593	*12593
Other kerosene	201	*175	*131	*88	83	*83
Bitumen	..	..	..	..	52	..
Total input	16810	*20826	*21123	*21038	23929	*24482
Total production	5418	6520	6634	6576	7177	8178
Estimated efficiency (% of production to input)	32	31	31	31	30	33

Statistics on electricity

New Zealand

Item	2009	2010	2011	2012	2013	2014
Production, trade and consumption	Million kilowatt-hours					
Total main activity and autoproducer	43453	44876	44465	44260	43265	43553
Combustible fuels	12834	12583	11233	13046	11739	9694
Hydro	24221	24716	25114	22901	23044	24336
Nuclear	..	..	..	..	..	..
Other	6398	7577	8118	8313	8482	9523
Main activity	42138	43487	43189	42886	41856	42155
Combustible fuels	11642	11321	10071	11785	10442	8417
Hydro	24209	24704	25102	22888	23031	24323
Nuclear	..	..	..	..	..	..
Other	6287	7462	8016	8213	8383	9415
Own use in electricity, CHP and heat plants	1373	1419	1357	1454	1383	1314
Net production	42080	43457	43108	42806	41882	42239
Imports	..	..	..	..	..	..
Exports	..	..	..	..	..	..
Losses	2992	3101	2999	2996	2888	2847
Consumption	38644	39703	39194	39125	38837	39231
Energy industries own use	512	548	593	611	613	595
By industry and construction	13609	14726	14709	13927	13817	13875
By transport	57	62	62	62	62	62
By households and other cons.	24466	24367	23830	24525	24345	24699
Net installed capacity	Thousand kilowatts					
Total main activity and autoproducer	9404	9457	9765	9619	9463	9704
Combustible fuels	2929	2929	3138	2988	2738	2741
Hydro	5326	5254	5254	5254	5263	5263
Nuclear	..	..	..	..	..	..
Other	1149	1274	1373	1377	1462	1700
Main activity	9151	9206	9514	9364	9205	9434
Combustible fuels	2706	2708	2917	2767	2517	2520
Hydro	5323	5251	5251	5251	5260	5260
Nuclear	..	..	..	..	..	..
Other	1122	1247	1346	1346	1428	1654
Combustible fuel input	Terajoules					
Brown Coal	25849	13111	15559	27170	16506	12399
Biogas	2498	2534	2720	2654	2634	2648
Gas-diesel oil	86	0	0	43	43	0
Natural gas	71231	81987	70248	71157	71128	58895
Coke-oven gas	1907	2181	1836	1846	2130	2007
Blast furnace gas	4857	5591	4921	5530	5541	5482
Fuelwood	3858	3976	3997	3860	3961	3961
Total input	110286	109380	99281	112260	101943	85392
Total production	46202	45299	40439	46966	42260	34898
Estimated efficiency (% of production to input)	42	41	41	42	41	41

Statistics on electricity

Nicaragua

Item	2009	2010	2011	2012	2013	2014
Production, trade and consumption	**Million kilowatt-hours**					
Total main activity and autoproducer	3453	3659	3824	4031	4163	4444
Combustible fuels	2749	2690	2896	2759	2466	2541
Hydro	297	504	444	419	456	395
Nuclear	..	..	..	..	..	..
Other	407	465	484	853	1241	1508
Main activity	3048	3230	3405	3529	3628	3899
Combustible fuels	2344	2261	2477	2258	1932	1997
Hydro	297	504	444	418	455	394
Nuclear	..	..	..	..	..	..
Other	407	465	484	853	1241	1508
Own use in electricity, CHP and heat plants	227	299	306	352	400	161
Net production	3226	3360	3518	3679	3763	4283
Imports	2	10	10	20	52	22
Exports	2	43	41	3	16	49
Losses	710	676	674	642	643	926
Consumption	2520	2641	2821	3041	3196	3325
Energy industries own use	..	..	..	..	..	..
By industry and construction	792	928	1010	1106	1065	1105
By transport	..	..	..	..	..	..
By households and other cons.	1728	1713	1811	1935	2131	2220
Net installed capacity	**Thousand kilowatts**					
Total main activity and autoproducer	983	1074	1110	1288	1292	1329
Combustible fuels	749	817	853	870	869	867
Hydro	106	106	106	107	121	121
Nuclear	..	..	..	..	..	..
Other	128	151	151	311	302	341
Main activity	860	951	987	1153	1157	1194
Combustible fuels	627	695	731	736	735	733
Hydro	105	105	105	106	120	120
Nuclear	..	..	..	..	..	..
Other	128	151	151	311	302	341
Combustible fuel input	**Terajoules**					
Gas-diesel oil	817	559	602	602	473	645
Fuel oil	21533	20968	22422	20362	17655	17978
Bagasse	*10090	*9775	*9523	*11670	*15269	15695
Vegetal waste	*1429	*1245	*2196	*1562	*1929	*3007
Total input	33869	32547	34743	34196	35326	37325
Total production	9896	9684	10426	9932	8878	9148
Estimated efficiency (% of production to input)	29	30	30	29	25	25

Statistics on electricity

Niger

Item	2009	2010	2011	2012	2013	2014
Production, trade and consumption	Million kilowatt-hours					
Total main activity and autoproducer	254	292	324	392	438	439
Combustible fuels	254	292	324	392	438	439
Hydro	..	..	..	..	..	..
Nuclear	..	..	..	..	..	..
Other	0	0	0	0	0	0
Main activity	234	269	270	338	399	383
Combustible fuels	234	269	270	338	399	383
Hydro	..	..	..	..	..	..
Nuclear	..	..	..	..	..	..
Other	..	..	..	..	..	..
Own use in electricity, CHP and heat plants	37	42	42	45	44	51
Net production	217	250	282	347	394	388
Imports	535	551	607	637	602	574
Exports	2	2	4	4	3	0
Losses	115	114	138	157	149	*185
Consumption	634	685	747	826	836	*797
Energy industries own use	..	..	..	..	..	..
By industry and construction	192	196	225	238	225	*215
By transport	..	..	..	..	..	..
By households and other cons.	442	489	521	588	611	*583
Net installed capacity	Thousand kilowatts					
Total main activity and autoproducer	135	*135	*135	*135	176	176
Combustible fuels	134	*134	*134	*134	172	172
Hydro	..	..	..	..	..	..
Nuclear	..	..	..	..	..	..
Other	1	*1	*1	*1	4	4
Main activity	105	*105	*105	*105	134	134
Combustible fuels	105	*105	*105	*105	134	134
Hydro	..	..	..	..	..	..
Nuclear	..	..	..	..	..	..
Other	..	..	..	..	..	..
Combustible fuel input	Terajoules					
Hard Coal	2765	3458	3858	4046	3597	3733
Gas-diesel oil	645	899	980	1441	2060	*1978
Fuel oil	133	137	0	0	0	0
Total input	3543	4494	4838	5487	5657	5711
Total production	914	1051	1166	1410	1576	1580
Estimated efficiency (% of production to input)	26	23	24	26	28	28

Statistics on electricity

Nigeria

Item	2009	2010	2011	2012	2013	2014
Production, trade and consumption	Million kilowatt-hours					
Total main activity and autoproducer	19777	26121	27034	28706	28883	30390
Combustible fuels	15248	19747	21151	23047	23557	25044
Hydro	4529	6374	5883	5659	5326	5346
Nuclear	..	..	..	..	..	..
Other	..	..	..	..	..	..
Main activity	14655	19356	20032	21271	23626	24808
Combustible fuels	10218	13112	14269	15727	18409	19571
Hydro	4437	6244	5763	5544	5217	5237
Nuclear	..	..	..	..	..	..
Other	..	..	..	..	..	..
Own use in electricity, CHP and heat plants	566	748	774	822	827	870
Net production	19211	25373	26260	27884	28056	29520
Imports	..	..	..	..	..	..
Exports	..	..	..	..	..	..
Losses	1160	4497	2581	2485	4367	4895
Consumption	18051	20876	23679	25399	23689	24625
Energy industries own use	140	216	232	240	202	186
By industry and construction	3109	3249	3699	3983	3899	4057
By transport	..	..	..	..	..	..
By households and other cons.	14802	17411	19748	21176	19588	20382
Net installed capacity	Thousand kilowatts					
Total main activity and autoproducer	*8694	*8417	*8902	*8902	*9822	*10396
Combustible fuels	*6764	*6487	*6972	*6972	*7892	*8418
Hydro	*1930	*1930	*1930	*1930	*1930	*1978
Nuclear	..	..	..	..	..	..
Other	..	..	..	..	..	..
Main activity	*6625	*6431	*6770	*6770	*7690	*7356
Combustible fuels	*4735	*4541	*4880	*4880	*5800	*5418
Hydro	*1890	*1890	*1890	*1890	*1890	*1938
Nuclear	..	..	..	..	..	..
Other	..	..	..	..	..	..
Combustible fuel input	Terajoules					
Natural gas	152480	197470	211510	230470	235570	250440
Total input	152480	197470	211510	230470	235570	250440
Total production	54893	71089	76144	82969	84805	90158
Estimated efficiency (% of production to input)	36	36	36	36	36	36

Statistics on electricity

Niue

Item	2009	2010	2011	2012	2013	2014
Production, trade and consumption	**Million kilowatt-hours**					
Total main activity and autoproducer	3	3	3	3	3	3
Combustible fuels	3	3	3	3	3	3
Hydro	..	..	..	..	..	..
Nuclear	..	..	..	..	..	..
Other	0	0	0	0	0	0
Main activity	3	3	3	3	3	3
Combustible fuels	3	3	3	3	3	3
Hydro	..	..	..	..	..	..
Nuclear	..	..	..	..	..	..
Other	0	0	0	0	0	0
Own use in electricity, CHP and heat plants	0	0	0	0	0	0
Net production	3	3	3	3	3	3
Imports	..	..	..	..	..	..
Exports	..	..	..	..	..	..
Losses	0	0	0	0	0	0
Consumption	3	3	3	3	3	3
Energy industries own use	..	..	..	..	..	..
By industry and construction	1	1	1	1	1	1
By transport	..	..	..	..	..	..
By households and other cons.	2	2	2	2	2	2
Net installed capacity	**Thousand kilowatts**					
Total main activity and autoproducer	2	2	2	2	2	2
Combustible fuels	2	2	2	2	2	2
Hydro	..	..	..	..	..	..
Nuclear	..	..	..	..	..	..
Other	0	0	0	0	0	0
Main activity	2	2	2	2	2	2
Combustible fuels	2	2	2	2	2	2
Hydro	..	..	..	..	..	..
Nuclear	..	..	..	..	..	..
Other	0	0	0	0	0	0
Combustible fuel input	**Terajoules**					
Gas-diesel oil	29	30	32	30	31	31
Total input	29	30	32	30	31	31
Total production	10	11	12	12	12	11
Estimated efficiency (% of production to input)	36	37	37	40	38	36

Northern Mariana Islands

Item	2009	2010	2011	2012	2013	2014
Production, trade and consumption	**Million kilowatt-hours**					
Total main activity and autoproducer	*397	*400	*407	*411	*415	*419
Combustible fuels	*397	*400	*407	*411	*415	*419
Hydro	..	..	..	..	..	..
Nuclear	..	..	..	..	..	..
Other	..	..	..	..	..	..
Main activity	*397	*400	*407	*411	*415	*419
Combustible fuels	*397	*400	*407	*411	*415	*419
Hydro	..	..	..	..	..	..
Nuclear	..	..	..	..	..	..
Other	..	..	..	..	..	..
Own use in electricity, CHP and heat plants	0	0	0	0	0	0
Net production	*397	*400	*407	*411	*415	*419
Imports	..	..	..	..	..	..
Exports	..	..	..	..	..	..
Losses	*89	*90	*91	*91	*92	*93
Consumption	*308	*310	*316	*319	*322	*326
Energy industries own use	..	..	..	..	..	..
By industry and construction	..	..	..	..	..	..
By transport	..	..	..	..	..	..
By households and other cons.	*308	*310	*316	*319	*322	*326
Net installed capacity	**Thousand kilowatts**					
Total main activity and autoproducer	*71	*74	*82	*88	*91	95
Combustible fuels	*71	*74	*82	*88	*91	95
Hydro	..	..	..	..	..	..
Nuclear	..	..	..	..	..	..
Other	..	..	..	..	..	..
Main activity	*71	*74	*82	*88	*91	95
Combustible fuels	*71	*74	*82	*88	*91	95
Hydro	..	..	..	..	..	..
Nuclear	..	..	..	..	..	..
Other	..	..	..	..	..	..
Combustible fuel input	**Terajoules**					
Total input	..	..	..	..	..	..
Total production	*1429	*1440	*1464	*1479	*1493	*1508
Estimated efficiency (% of production to input)	..	..	..	..	..	..

Statistics on electricity

Norway

Item	2009	2010	2011	2012	2013	2014
Production, trade and consumption	Million kilowatt-hours					
Total main activity and autoproducer	131773	123640	127655	147716	133975	142327
Combustible fuels	4644	5454	4739	3354	3152	3172
Hydro	126077	117152	121552	142812	128699	136636
Nuclear	..	..	..	..	..	..
Other	1052	1034	1364	1550	2124	2519
Main activity	122260	113590	118117	139097	125503	134656
Combustible fuels	208	303	360	457	482	521
Hydro	121070	112379	116451	137092	123140	131919
Nuclear	..	..	..	..	..	..
Other	982	908	1306	1548	1881	2216
Own use in electricity, CHP and heat plants	608	569	593	560	620	707
Net production	131165	123071	127062	147156	133355	141620
Imports	5651	14673	11255	4190	10135	6347
Exports	14634	7124	14329	22006	15141	21932
Losses	7578	9491	8109	9098	8039	8586
Consumption	113709	120283	115215	119390	119310	116582
Energy industries own use	6422	6832	7826	8612	7611	8173
By industry and construction	41289	44544	44043	43489	43342	45334
By transport	660	690	689	768	730	762
By households and other cons.	65338	68217	62657	66521	67627	62313
Net installed capacity	Thousand kilowatts					
Total main activity and autoproducer	31255	31688	32082	32860	33486	33651
Combustible fuels	1258	1535	1566	1611	1600	1600
Hydro	29539	29693	29969	30509	31033	31153
Nuclear	..	..	..	..	..	..
Other	458	460	547	740	853	898
Main activity	29616	29614	30008	30786	31412	31577
Combustible fuels	529	542	573	618	607	607
Hydro	28651	28634	28910	29450	29974	30094
Nuclear	..	..	..	..	..	..
Other	436	438	525	718	831	876
Combustible fuel input	Terajoules					
Hard Coal	731	731	703	618	703	703
Biogas	85	88	82	81	93	93
Gas-diesel oil	129	129	129	172	129	172
Refinery gas	0	0	0	0	0	0
Natural gas	25481	36196	27350	18408	15058	16098
Blast furnace gas	292	329	523	658	507	610
Fuelwood	866	1271	1286	1350	1120	80
Municipal waste	3146	6880	8910	9500	11647	12700
Industrial waste	60	115	115	295	112	103
Other liquid biofuels	0	0	0	0	0	0
Others	0	0	0	0	0	0
Total input	30790	45739	39097	31082	29368	30559
Total production	16718	19634	17060	12074	11347	11419
Estimated efficiency (% of production to input)	54	43	44	39	39	37

Statistics on electricity

Oman

Item	2009	2010	2011	2012	2013	2014
Production, trade and consumption	**Million kilowatt-hours**					
Total main activity and autoproducer	18445	19819	21874	25017	26240	29128
Combustible fuels	18445	19819	21874	25017	26240	29128
Hydro	..	..	..	..	..	..
Nuclear	..	..	..	..	..	..
Other	..	..	..	..	..	..
Main activity	18445	19819	21874	25017	26240	29128
Combustible fuels	18445	19819	21874	25017	26240	29128
Hydro	..	..	..	..	..	..
Nuclear	..	..	..	..	..	..
Other	..	..	..	..	..	..
Own use in electricity, CHP and heat plants	622	659	520	652	579	785
Net production	17823	19160	21354	24365	25661	28343
Imports	..	..	..	..	..	..
Exports	..	..	..	..	..	..
Losses	3340	3027	2842	3410	2871	3170
Consumption	14483	16134	18513	20956	22790	25173
Energy industries own use	..	..	..	..	..	..
By industry and construction	1175	1541	2584	3436	3686	4189
By transport	..	..	..	..	..	..
By households and other cons.	13308	14593	15929	17520	19104	20984
Net installed capacity	**Thousand kilowatts**					
Total main activity and autoproducer	4202	4265	4861	5808	6598	8214
Combustible fuels	4202	4265	4861	5808	6598	8214
Hydro	..	..	..	..	..	..
Nuclear	..	..	..	..	..	..
Other	..	..	..	..	..	..
Main activity	4202	4265	4861	5808	6598	8214
Combustible fuels	4202	4265	4861	5808	6598	8214
Hydro	..	..	..	..	..	..
Nuclear	..	..	..	..	..	..
Other	..	..	..	..	..	..
Combustible fuel input	**Terajoules**					
Gas-diesel oil	4601	5461	5676	6278	7009	7697
Natural gas	229758	242699	257745	289108	285035	305274
Total input	234359	248160	263421	295386	292044	312971
Total production	66402	71348	78746	90061	94464	104861
Estimated efficiency (% of production to input)	28	29	30	30	32	34

Statistics on electricity

Pakistan

Item	2009	2010	2011	2012	2013	2014
Production, trade and consumption	Million kilowatt-hours					
Total main activity and autoproducer	95358	94383	95090	*96126	*104076	*105305
Combustible fuels	64371	59152	61308	*61716	*66716	*68390
Hydro	28093	31811	28517	29857	31873	31428
Nuclear	2894	3420	5265	4553	5090	5090
Other	..	..	..	..	397	397
Main activity	58544	57169	55416	55952	60902	61048
Combustible fuels	27557	21938	21634	*21542	*23542	*24133
Hydro	28093	31811	28517	29857	31873	31428
Nuclear	2894	3420	5265	4553	5090	5090
Other	..	..	..	..	397	397
Own use in electricity, CHP and heat plants	2303	2239	2550	3341	3757	3801
Net production	93055	92144	92540	*92785	*100319	*101504
Imports	249	269	274	375	419	428
Exports	..	..	..	..	..	..
Losses	18957	15315	16054	16372	16932	18333
Consumption	74347	77099	76761	76788	83409	86328
Energy industries own use	..	..	..	..	..	..
By industry and construction	19823	21207	21801	22313	24356	25510
By transport	2	1	1	0	0	0
By households and other cons.	54522	55891	54959	54475	59053	60818
Net installed capacity	Thousand kilowatts					
Total main activity and autoproducer	20921	22477	22797	*22812	*23686	*23686
Combustible fuels	13978	15209	15454	*15289	*15888	*15888
Hydro	6481	6481	6556	6773	6893	*6893
Nuclear	462	787	787	750	750	*750
Other	..	..	..	..	*155	*155
Main activity	13798	14114	14444	14629	15128	*15128
Combustible fuels	6855	6846	7101	*7106	*7330	*7330
Hydro	6481	6481	6556	6773	6893	*6893
Nuclear	462	787	787	750	750	*750
Other	..	..	..	..	*155	*155
Combustible fuel input	Terajoules					
Hard Coal	3140	2394	2625	1545	3949	3980
Gas-diesel oil	10737	4300	8308	8944	12470	13072
Fuel oil	346010	324776	299021	*305626	*352975	*369620
Natural gas	314128	287024	297593	313145	291850	289406
Total input	674015	618494	607546	629260	*661243	*676077
Total production	231736	212947	220709	*222178	*240178	*246204
Estimated efficiency (% of production to input)	34	34	36	35	36	36

157

Statistics on electricity

Palau

Item	2009	2010	2011	2012	2013	2014
Production, trade and consumption	**Million kilowatt-hours**					
Total main activity and autoproducer	*83	*80	*78	*78	*78	*79
Combustible fuels	*83	*80	*78	*78	*78	*79
Hydro	..	..	..	..	..	..
Nuclear	..	..	..	..	..	..
Other	..	..	..	..	..	..
Main activity	*77	*74	*72	*72	*72	*73
Combustible fuels	*77	*74	*72	*72	*72	*73
Hydro	..	..	..	..	..	..
Nuclear	..	..	..	..	..	..
Other	..	..	..	..	..	..
Own use in electricity, CHP and heat plants	*5	*5	*5	*5	*5	*5
Net production	*78	*75	*73	*73	*73	*74
Imports	..	..	..	..	..	..
Exports	..	..	..	..	..	..
Losses	*12	*11	*11	*11	*11	*11
Consumption	69	72	71	*67	*64	*65
Energy industries own use	..	..	..	..	..	..
By industry and construction	3	3	3	*3	*3	*3
By transport	..	..	..	..	..	..
By households and other cons.	66	69	68	*64	*61	*62
Net installed capacity	**Thousand kilowatts**					
Total main activity and autoproducer	*30	*30	*30	*30	*30	30
Combustible fuels	*30	*30	*30	*30	*30	30
Hydro	..	..	..	..	..	..
Nuclear	..	..	..	..	..	..
Other	..	..	..	..	..	..
Main activity	*28	*28	*28	*28	*28	28
Combustible fuels	*28	*28	*28	*28	*28	28
Hydro	..	..	..	..	..	..
Nuclear	..	..	..	..	..	..
Other	..	..	..	..	..	..
Combustible fuel input	**Terajoules**					
Gas-diesel oil	*963	*933	*890	*847	*912	*916
Total input	*963	*933	*890	*847	*912	*916
Total production	*298	*289	*279	*279	*279	*284
Estimated efficiency (% of production to input)	31	31	31	33	31	31

Statistics on electricity

Panama

Item	2009	2010	2011	2012	2013	2014
Production, trade and consumption	Million kilowatt-hours					
Total main activity and autoproducer	6909	7419	7857	8606	8962	9287
Combustible fuels	3012	3225	3759	3218	3808	4136
Hydro	3897	4194	4098	5388	5154	5034
Nuclear	..	..	..	..	..	..
Other	..	..	..	..	..	118
Main activity	6892	7397	7834	8580	8929	9256
Combustible fuels	2995	3203	3736	3192	3775	4105
Hydro	3897	4194	4098	5388	5154	5034
Nuclear	..	..	..	..	..	..
Other	..	..	..	..	..	118
Own use in electricity, CHP and heat plants	7	7	8	8	8	11
Net production	6902	7412	7849	8598	8954	9276
Imports	64	71	72	19	75	193
Exports	95	39	8	59	71	99
Losses	908	1080	1080	1177	1206	1328
Consumption	6018	6388	6695	7176	7511	7832
Energy industries own use	..	..	..	..	..	..
By industry and construction	570	651	660	695	706	689
By transport	..	..	..	..	..	19
By households and other cons.	5448	5737	6035	6481	6805	7124
Net installed capacity	Thousand kilowatts					
Total main activity and autoproducer	1819	1976	2391	2422	2546	2829
Combustible fuels	940	1040	1040	954	1032	1148
Hydro	879	936	1351	1468	1494	1623
Nuclear	..	..	..	..	..	..
Other	..	0	0	0	20	57
Main activity	1603	1752	2167	2198	2333	2560
Combustible fuels	785	881	881	795	879	939
Hydro	818	871	1286	1403	1434	1563
Nuclear	..	..	..	..	..	..
Other	..	0	0	0	20	57
Combustible fuel input	Terajoules					
Hard Coal	..	..	6115	8333	8798	9133
Gas-diesel oil	5203	8084	10062	3010	5676	9546
Fuel oil	22503	21695	20038	18301	12484	6828
Bagasse	1992	2189	2221	2476	3081	2874
Total input	29698	31968	38436	32121	30038	28381
Total production	10843	11610	13534	11583	13708	14888
Estimated efficiency (% of production to input)	37	36	35	36	46	52

Statistics on electricity

Papua New Guinea

Item	2009	2010	2011	2012	2013	2014
Production, trade and consumption	**Million kilowatt-hours**					
Total main activity and autoproducer	3486	3631	3666	3764	4169	*4169
Combustible fuels	2120	2252	2275	2359	2752	*2752
Hydro	944	957	969	983	995	*995
Nuclear	..	..	..	..	..	..
Other	422	422	422	422	422	*422
Main activity	866	876	885	896	975	*975
Combustible fuels	273	270	267	264	340	*340
Hydro	593	606	618	632	635	*635
Nuclear	..	..	..	..	..	..
Other	..	..	..	..	..	..
Own use in electricity, CHP and heat plants	0	1	0	1	0	0
Net production	3486	3630	3666	3763	4169	*4169
Imports	..	..	..	..	..	..
Exports	..	..	..	..	..	..
Losses	*383	*406	*407	*434	242	*158
Consumption	3103	3224	3259	3329	3926	4011
Energy industries own use	..	..	..	..	..	..
By industry and construction	2268	2362	2362	2432	2985	3056
By transport	..	..	..	..	..	..
By households and other cons.	*835	*862	*897	*897	*942	955
Net installed capacity	**Thousand kilowatts**					
Total main activity and autoproducer	*763	*788	*788	*795	*865	*975
Combustible fuels	*469	*478	*478	*478	*548	*650
Hydro	*238	254	254	261	261	*269
Nuclear	..	..	..	..	..	..
Other	56	56	56	56	56	*56
Main activity	*272	273	273	280	280	*320
Combustible fuels	*109	118	118	*118	*118	*150
Hydro	*163	155	155	162	162	*170
Nuclear	..	..	..	..	..	..
Other	..	..	..	..	..	..
Combustible fuel input	**Terajoules**					
Gas-diesel oil	11180	10320	10965	4312	8625	*9000
Fuel oil	12605	12524	12928	17920	19175	*19000
Natural gas	5658	5658	5896	5024	5150	*5000
Total input	29443	28502	29789	27256	32950	*33000
Total production	7632	8107	8190	8492	9907	*9907
Estimated efficiency (% of production to input)	26	28	27	31	30	30

Statistics on electricity

Paraguay

Item	2009	2010	2011	2012	2013	2014
Production, trade and consumption	Million kilowatt-hours					
Total main activity and autoproducer	54940	54066	57625	60235	60381	55282
Combustible fuels	1	1	1	3	3	6
Hydro	54939	54065	57624	60232	60378	55276
Nuclear	..	..	..	..	..	..
Other	..	..	..	..	..	..
Main activity	54940	54066	57625	60235	60381	55282
Combustible fuels	1	1	1	3	3	6
Hydro	54939	54065	57624	60232	60378	55276
Nuclear	..	..	..	..	..	..
Other	..	..	..	..	..	..
Own use in electricity, CHP and heat plants	308	449	501	504	506	450
Net production	54631	53617	57124	59730	59875	54832
Imports	..	..	..	..	..	..
Exports	45042	43378	46120	47663	47365	41400
Losses	3130	3369	3433	3845	3492	3637
Consumption	6451	6870	7571	8222	9018	9795
Energy industries own use	..	..	..	..	..	..
By industry and construction	2735	2889	3114	3343	3567	3801
By transport	..	..	..	..	..	..
By households and other cons.	3716	3980	4456	4879	5451	5993
Net installed capacity	Thousand kilowatts					
Total main activity and autoproducer	8816	8816	8816	8816	8825	8825
Combustible fuels	6	6	6	6	15	15
Hydro	8810	8810	8810	8810	8810	8810
Nuclear	..	..	..	..	..	..
Other	..	..	..	..	..	..
Main activity	8816	8816	8816	8816	8825	8825
Combustible fuels	6	6	6	6	15	15
Hydro	8810	8810	8810	8810	8810	8810
Nuclear	..	..	..	..	..	..
Other	..	..	..	..	..	..
Combustible fuel input	Terajoules					
Gas-diesel oil	4	5	6	22	22	57
Total input	4	5	6	22	22	57
Total production	3	3	3	11	11	21
Estimated efficiency (% of production to input)	62	62	58	51	51	37

161

Statistics on electricity

Peru

Item	2009	2010	2011	2012	2013	2014
Production, trade and consumption	**Million kilowatt-hours**					
Total main activity and autoproducer	32929	35890	38811	40044	43330	45508
Combustible fuels	13437	15847	17248	17951	20812	22882
Hydro	19491	20042	21562	22032	22320	22171
Nuclear	..	..	..	..	..	..
Other	1	1	1	61	198	455
Main activity	30907	33529	36253	37369	40665	42844
Combustible fuels	11899	13970	15220	15817	18757	20778
Hydro	19007	19558	21032	21491	21709	21611
Nuclear	..	..	..	..	..	..
Other	1	1	1	61	198	455
Own use in electricity, CHP and heat plants	499	563	558	564	589	800
Net production	32430	35327	38253	39480	42741	44708
Imports	0	0	6	5	0	0
Exports	63	112	0	2	0	0
Losses	2686	3634	3109	3380	4556	5018
Consumption	29681	31582	35125	35969	38208	39719
Energy industries own use	..	..	..	..	..	..
By industry and construction	15538	16789	18580	19027	20686	21798
By transport	0	0	3	3	22	37
By households and other cons.	14143	14793	16542	16939	17499	17884
Net installed capacity	**Thousand kilowatts**					
Total main activity and autoproducer	7983	8614	8691	9699	11051	11203
Combustible fuels	4709	5175	5240	6134	7414	7302
Hydro	3273	3438	3451	3484	3556	3662
Nuclear	..	..	..	..	..	..
Other	1	1	1	81	81	239
Main activity	6724	7310	7314	8267	9635	9739
Combustible fuels	3540	3964	3957	4806	6104	5942
Hydro	3183	3345	3357	3381	3451	3558
Nuclear	..	..	..	..	..	..
Other	1	1	1	81	81	239
Combustible fuel input	**Terajoules**					
Hard Coal	9225	10250	10385	8416	11437	7016
Biogas	0	0	0	877	806	821
Gas-diesel oil	7353	7095	7009	7955	5633	5074
Fuel oil	13776	18988	21897	8201	8565	6181
Refinery gas	198	198	1238	1238	1287	1485
Natural gas	114228	145576	149954	164487	150325	175586
Bagasse	..	..	12502	10324	15622	21452
Vegetal waste	6622	8358	..	..	..	..
Biodiesel	110	258	0	0	0	0
Total input	151513	190723	202985	201498	193675	217615
Total production	48373	57049	62092	64624	74925	82375
Estimated efficiency (% of production to input)	32	30	31	32	39	38

Statistics on electricity

Philippines

Item	2009	2010	2011	2012	2013	2014
Production, trade and consumption	**Million kilowatt-hours**					
Total main activity and autoproducer	61968	67777	69210	72956	75300	77295
Combustible fuels	41774	49964	49463	52360	55592	57664
Hydro	9805	7820	9715	10269	10036	9154
Nuclear	..	..	..	..	..	..
Other	10389	9992	10032	10327	9672	10477
Main activity	61934	67743	69176	72922	75266	77261
Combustible fuels	41757	49947	49446	52343	55575	57647
Hydro	9788	7803	9698	10252	10019	9137
Nuclear	..	..	..	..	..	..
Other	10389	9992	10032	10327	9672	10477
Own use in electricity, CHP and heat plants	3524	4677	5399	5351	5545	6680
Net production	58444	63100	63811	67605	69755	70615
Imports	..	..	..	..	..	..
Exports	..	..	..	..	..	..
Losses	7542	7800	7680	8360	7741	7270
Consumption	50868	55266	56098	59211	61566	63345
Energy industries own use	..	..	..	..	..	..
By industry and construction	17084	18576	19334	20071	20677	21429
By transport	111	109	111	117	112	111
By households and other cons.	33673	36581	36653	39023	40777	41805
Net installed capacity	**Thousand kilowatts**					
Total main activity and autoproducer	15635	16386	16189	17052	17351	17970
Combustible fuels	10311	10931	10782	11514	11793	12056
Hydro	3307	3416	3507	3537	3537	3559
Nuclear	..	..	..	..	..	..
Other	2017	2039	1900	2001	2021	2355
Main activity	15609	16360	16163	17026	17325	17944
Combustible fuels	10301	10921	10772	11504	11783	12046
Hydro	3291	3400	3491	3521	3521	3543
Nuclear	..	..	..	..	..	..
Other	2017	2039	1900	2001	2021	2355
Combustible fuel input	**Terajoules**					
Brown Coal	185988	213093	242221	263790	326843	344453
Gas-diesel oil	11507	14573	7994	8544	10913	14732
Fuel oil	38764	48617	25973	32869	35055	44137
Natural gas	146914	137502	147207	141006	128778	135138
Bagasse	..	180	998	2035	2122	1818
Municipal waste	343	361	1095	926	1499	1639
Biodiesel	204	256	55	152	193	269
Total input	383720	414582	425543	449323	505403	542186
Total production	150387	179871	178068	188497	200131	207591
Estimated efficiency (% of production to input)	39	43	42	42	40	38

Statistics on electricity

Poland

Item	2009	2010	2011	2012	2013	2014
Production, trade and consumption		Million kilowatt-hours				
Total main activity and autoproducer	151720	157657	163548	162139	164557	159059
Combustible fuels	147669	152422	157477	154845	155444	148500
Hydro	2974	3488	2761	2465	2997	2734
Nuclear	..	..	..	..	..	..
Other	1077	1747	3310	4829	6116	7825
Main activity	144838	149691	155206	153690	155640	150178
Combustible fuels	140789	144541	149242	146480	146642	139771
Hydro	2972	3486	2759	2463	2994	2731
Nuclear	..	..	..	..	..	..
Other	1077	1664	3205	4747	6004	7676
Own use in electricity, CHP and heat plants	13815	14200	14635	14490	14514	13845
Net production	137905	143457	148913	147649	150043	145214
Imports	7403	6310	6780	9803	7801	13508
Exports	9594	7664	12022	12643	12322	11342
Losses	12533	11851	10638	10884	10247	10250
Consumption	123181	130252	133033	133925	135275	137130
Energy industries own use	10489	11189	11058	11251	11216	11267
By industry and construction	39776	41825	44615	45219	47828	48059
By transport	3220	3338	3317	3200	3156	3007
By households and other cons.	69696	73900	74043	74255	73075	74797
Net installed capacity		Thousand kilowatts				
Total main activity and autoproducer	33032	33360	34554	35283	35815	35989
Combustible fuels	29983	29908	30405	30365	30027	29760
Hydro	2338	2342	2346	2351	2355	2364
Nuclear	..	..	..	..	..	..
Other	711	1110	1803	2567	3433	3865
Main activity	31375	31564	32823	33514	34009	34141
Combustible fuels	28329	28115	28678	28600	28226	27942
Hydro	2337	2341	2345	2350	2354	2363
Nuclear	..	..	..	..	..	..
Other	709	1108	1800	2564	3429	3836
Combustible fuel input		Terajoules				
Hard Coal	992541	1065525	1045664	958319	978720	900363
Brown Coal	500305	476989	516512	524934	542388	514876
Biogas	3096	2763	3312	4204	4872	5718
Fuel oil	18907	20927	16402	14665	12726	11352
Natural gas	45628	45252	53135	57640	47363	47736
Coke-oven gas	18525	20215	17958	17309	19410	19325
Blast furnace gas	7443	9954	11001	11328	11729	13937
Fuelwood	*40896	*47153	*56619	65251	*64200	*73992
Vegetal waste	*12251	*17088	*21502	*39850	*21400	*21002
Other recovered gases	4813	5026	5332	5202	5247	5057
Others	2873	3540	3096	2869	3076	2374
Total input	1647278	1714433	1750534	1701570	1711131	1615733
Total production	531608	548719	566917	557442	559598	534600
Estimated efficiency (% of production to input)	32	32	32	33	33	33

Statistics on electricity

Portugal

Item	2009	2010	2011	2012	2013	2014
Production, trade and consumption	Million kilowatt-hours					
Total main activity and autoproducer	50207	54090	52463	46614	51672	52802
Combustible fuels	33277	27950	30693	29151	24110	23440
Hydro	9009	16547	12115	6660	14868	16412
Nuclear	..	..	..	..	..	..
Other	7921	9593	9655	10803	12694	12950
Main activity	43693	46348	44116	38270	42662	44367
Combustible fuels	26779	20230	22442	20967	15330	15323
Hydro	8993	16528	12101	6650	14848	16389
Nuclear	..	..	..	..	..	..
Other	7921	9590	9573	10653	12484	12655
Own use in electricity, CHP and heat plants	1489	1308	1335	1361	1261	1276
Net production	48718	52782	51128	45253	50411	51526
Imports	7598	5814	6742	10766	8100	7247
Exports	2822	3191	3929	2871	5324	6344
Losses	3792	4280	4090	4707	5455	5209
Consumption	49702	51125	49851	48441	47732	47220
Energy industries own use	1847	1237	1487	2200	2475	2025
By industry and construction	16173	17468	16957	15932	15993	15395
By transport	483	475	404	401	301	302
By households and other cons.	31199	31945	31003	29908	28963	29498
Net installed capacity	Thousand kilowatts					
Total main activity and autoproducer	17403	18932	19924	19747	18900	19125
Combustible fuels	8846	9871	9936	9360	8308	8113
Hydro	5091	5106	5535	5712	5661	5715
Nuclear	..	..	..	..	..	..
Other	3466	3955	4453	4675	4931	5297
Main activity	15718	17106	17969	17635	16841	17020
Combustible fuels	7167	8051	8051	7357	6409	6205
Hydro	5085	5100	5529	5706	5655	5709
Nuclear	..	..	..	..	..	..
Other	3466	3955	4389	4572	4777	5106
Combustible fuel input	Terajoules					
Hard Coal	118764	66949	92278	121920	110009	111645
Biogas	996	1287	1883	2362	2736	3078
Gas-diesel oil	946	1161	860	860	860	903
Fuel oil	29856	23109	19796	15958	11393	9494
Refinery gas	941	1485	743	941	1188	941
Natural gas	119662	129071	127891	98507	79197	72602
Vegetal waste	*8058	*8115	*8748	*8891	*8996	*12376
Municipal waste	8293	8030	8245	7197	8097	6845
Industrial waste	83	343	216	127	101	91
Biodiesel	..	..	..	..	37	37
Total input	287598	239550	260659	256763	222613	218011
Total production	119797	100620	110495	104944	86796	84384
Estimated efficiency (% of production to input)	42	42	42	41	39	39

Statistics on electricity

Puerto Rico

Item	2009	2010	2011	2012	2013	2014
Production, trade and consumption	**Million kilowatt-hours**					
Total main activity and autoproducer	23239	23497	22507	22556	21743	21317
Combustible fuels	23069	23365	22358	22429	21509	20969
Hydro	170	133	150	125	91	81
Nuclear	..	..	..	..	..	..
Other	..	..	..	2	144	267
Main activity	22989	23247	22257	22306	21493	21067
Combustible fuels	22819	23115	22108	22179	21259	20719
Hydro	170	133	150	125	91	81
Nuclear	..	..	..	..	..	..
Other	..	..	..	2	144	267
Own use in electricity, CHP and heat plants	1226	688	618	1102	484	795
Net production	22013	22809	21890	21454	21259	20522
Imports	..	..	..	..	..	..
Exports	..	..	..	..	..	..
Losses	*1789	*1855	*1663	*1622	*1303	*570
Consumption	19975	20704	19977	19582	19706	19702
Energy industries own use	..	..	..	..	..	..
By industry and construction	3290	3048	2882	2779	2579	2338
By transport	..	..	..	..	..	..
By households and other cons.	16685	17656	17095	16803	17127	17364
Net installed capacity	**Thousand kilowatts**					
Total main activity and autoproducer	5925	5901	5900	5903	6024	*6036
Combustible fuels	5825	5801	5800	5800	5800	*5800
Hydro	100	100	100	100	100	*100
Nuclear	..	..	..	..	..	..
Other	..	0	0	*3	124	*136
Main activity	5864	5840	5839	5842	5963	*5975
Combustible fuels	5764	5740	5739	5739	5739	*5739
Hydro	100	100	100	100	100	*100
Nuclear	..	..	..	..	..	..
Other	..	0	0	*3	124	*136
Combustible fuel input	**Terajoules**					
Hard Coal	38700	38700	38700	38700	32327	*32327
Fuel oil	153627	159931	145739	149522	142871	137909
Natural gas	29521	29995	29257	53059	60856	64855
Total input	221848	228626	213696	241281	236055	235091
Total production	83048	84112	80488	80743	77431	75489
Estimated efficiency (% of production to input)	37	37	38	33	33	32

Statistics on electricity

Qatar

Item	2009	2010	2011	2012	2013	2014
Production, trade and consumption	Million kilowatt-hours					
Total main activity and autoproducer	24158	28144	30730	34787	34668	38692
Combustible fuels	24158	28144	30730	34787	34668	38692
Hydro	..	..	..	..	..	..
Nuclear	..	..	..	..	..	..
Other	..	..	..	..	..	..
Main activity	3037	2422	2360	2988	3090	3502
Combustible fuels	3037	2422	2360	2988	3090	3502
Hydro	..	..	..	..	..	..
Nuclear	..	..	..	..	..	..
Other	..	..	..	..	..	..
Own use in electricity, CHP and heat plants	1510	1759	2347	2436	2444	2568
Net production	22648	26385	28383	32351	32224	36124
Imports	..	..	..	..	..	..
Exports	..	..	..	..	..	..
Losses	1517	1767	624	2168	2159	2341
Consumption	21131	24618	27759	30184	30066	33784
Energy industries own use	..	..	..	..	..	..
By industry and construction	6327	7773	9089	9798	9944	11568
By transport	..	..	..	..	..	..
By households and other cons.	14804	16845	18670	20386	20122	22216
Net installed capacity	Thousand kilowatts					
Total main activity and autoproducer	*6001	8751	8751	8751	8751	8751
Combustible fuels	*6001	8751	8751	8751	8751	8751
Hydro	..	..	..	..	..	..
Nuclear	..	..	..	..	..	..
Other	..	..	..	..	..	..
Main activity	*2190	2190	2190	2190	2190	2190
Combustible fuels	*2190	2190	2190	2190	2190	2190
Hydro	..	..	..	..	..	..
Nuclear	..	..	..	..	..	..
Other	..	..	..	..	..	..
Combustible fuel input	Terajoules					
Natural gas	243900	276270	299648	341440	442417	381028
Total input	243900	276270	299648	341440	442417	381028
Total production	86969	101318	110628	125233	124805	139291
Estimated efficiency (% of production to input)	36	37	37	37	28	37

Statistics on electricity

Republic of Moldova

Item	2009	2010	2011	2012	2013	2014
Production, trade and consumption	**Million kilowatt-hours**					
Total main activity and autoproducer	1033	1064	1016	932	905	963
Combustible fuels	978	985	940	898	859	902
Hydro	55	79	76	34	45	59
Nuclear	..	..	..	..	..	..
Other	0	0	0	0	1	2
Main activity	1012	1026	982	900	861	891
Combustible fuels	957	947	906	866	815	831
Hydro	55	79	76	34	45	59
Nuclear	..	..	..	..	..	..
Other	0	0	0	0	1	1
Own use in electricity, CHP and heat plants	146	141	136	129	123	127
Net production	887	923	880	803	782	836
Imports	2941	3033	3145	3279	3332	3342
Exports	..	..	..	..	..	..
Losses	450	468	453	446	431	481
Consumption	3378	3488	3571	3636	3683	3697
Energy industries own use	197	200	191	166	37	49
By industry and construction	688	790	815	845	873	905
By transport	50	46	50	54	59	47
By households and other cons.	2443	2452	2515	2571	2714	2696
Net installed capacity	**Thousand kilowatts**					
Total main activity and autoproducer	504	502	503	487	488	421
Combustible fuels	488	486	487	471	471	403
Hydro	16	16	16	16	16	16
Nuclear	..	..	..	..	..	..
Other	..	0	0	0	1	2
Main activity	346	346	346	346	347	347
Combustible fuels	330	330	330	330	330	330
Hydro	16	16	16	16	16	16
Nuclear	..	..	..	..	..	..
Other	..	0	0	0	1	1
Combustible fuel input	**Terajoules**					
Hard Coal	0	0	0	0	0	0
Biogas	0	0	0	0	36	178
Gas-diesel oil	43	0	0	0	0	0
Fuel oil	929	808	606	485	566	485
Other oil products	0	0	0	0	0	0
Natural gas	28330	28827	28080	27698	27061	27931
Fuelwood	0	0	0	0	0	0
Vegetal waste	0	0	0	0	0	0
Industrial waste	0	0	0	0	0	0
Total input	29302	29635	28686	28183	27663	28594
Total production	3521	3546	3384	3233	3092	3247
Estimated efficiency (% of production to input)	12	12	12	11	11	11

Statistics on electricity

Réunion

Item	2009	2010	2011	2012	2013	2014
Production, trade and consumption	Million kilowatt-hours					
Total main activity and autoproducer	2618	2700	2750	2812	2813	2857
Combustible fuels	2052	2065	2195	2114	2017	2180
Hydro	531	542	401	488	557	426
Nuclear	..	..	..	..	..	..
Other	36	93	154	209	239	252
Main activity	2618	2700	2750	2812	2813	2857
Combustible fuels	2052	2065	2195	2114	2017	2180
Hydro	531	542	401	488	557	426
Nuclear	..	..	..	..	..	..
Other	36	93	154	209	239	252
Own use in electricity, CHP and heat plants	0	0	0	0	0	0
Net production	2618	2700	2750	2812	2813	2857
Imports	..	..	..	..	..	..
Exports	..	..	..	..	..	..
Losses	230	233	243	284	269	259
Consumption	2388	2467	2498	2540	2555	2598
Energy industries own use	..	..	..	..	..	..
By industry and construction	..	..	..	..	..	..
By transport	..	..	..	..	..	..
By households and other cons.	2388	2467	2498	2540	2555	2598
Net installed capacity	Thousand kilowatts					
Total main activity and autoproducer	649	709	766	868	875	834
Combustible fuels	471	471	477	567	564	510
Hydro	121	133	146	134	134	136
Nuclear	..	..	..	..	..	..
Other	57	104	144	167	178	188
Main activity	649	709	766	868	875	834
Combustible fuels	471	471	477	567	564	510
Hydro	121	133	146	134	134	136
Nuclear	..	..	..	..	..	..
Other	57	104	144	167	178	188
Combustible fuel input	Terajoules					
Hard Coal	16788	17923	17634	18313	17529	16494
Biogas	72	80	80	96	155	138
Gas-diesel oil	2047	1677	2365	1780	331	224
Fuel oil	3268	3050	3676	3123	3592	5612
Bagasse	4161	4191	4082	4095	3920	4145
Total input	26337	26922	27837	27407	25526	26612
Total production	7385	7434	7903	7610	7262	7847
Estimated efficiency (% of production to input)	28	28	28	28	28	29

Statistics on electricity

Romania

Item	2009	2010	2011	2012	2013	2014
Production, trade and consumption	**Million kilowatt-hours**					
Total main activity and autoproducer	58014	60979	62217	59045	58887	65676
Combustible fuels	30446	28807	34135	32594	27021	26903
Hydro	15807	20243	14946	12337	15308	19280
Nuclear	11752	11623	11747	11466	11618	11676
Other	9	306	1389	2648	4940	7817
Main activity	55709	58322	59437	54793	52368	59581
Combustible fuels	28362	26424	31509	28686	21103	21866
Hydro	15587	19975	14817	12179	15062	18953
Nuclear	11752	11623	11747	11466	11618	11676
Other	8	300	1364	2462	4585	7086
Own use in electricity, CHP and heat plants	5255	5056	5727	5351	4753	4977
Net production	52759	55923	56490	53694	54134	60699
Imports	651	767	3410	3903	2737	2811
Exports	2946	3041	5316	3650	4753	9937
Losses	7029	7058	7141	7062	7021	7097
Consumption	43020	46440	47533	47213	45408	46424
Energy industries own use	5413	5123	4819	4827	4780	4519
By industry and construction	18183	20381	21083	20405	18823	19855
By transport	1383	1355	1424	1228	1126	1062
By households and other cons.	18041	19581	20207	20753	20679	20988
Net installed capacity	**Thousand kilowatts**					
Total main activity and autoproducer	19551	19911	20499	21774	23033	24054
Combustible fuels	11675	11638	11616	11945	11393	11323
Hydro	6450	6474	6483	6548	6610	6613
Nuclear	1411	1411	1411	1411	1411	1411
Other	15	388	989	1870	3619	4707
Main activity	18736	19146	19734	19734	19974	20896
Combustible fuels	10944	10962	10948	10425	9064	8981
Hydro	6368	6391	6411	6455	6509	6523
Nuclear	1411	1411	1411	1411	1411	1411
Other	13	382	964	1443	2990	3981
Combustible fuel input	**Terajoules**					
Brown Coal	267055	249990	309994	276108	208366	207524
Biogas	5	8	88	183	422	597
Gas-diesel oil	129	215	215	258	344	215
Fuel oil	13170	8080	7676	4888	1535	889
Refinery gas	1139	4703	5792	5396	3020	2525
Other oil products	884	0	1849	2935	4020	4543
Natural gas	108541	104326	114919	116627	102871	95207
Blast furnace gas	366	868	315	235	449	690
Fuelwood	55	810	1366	1772	1947	5005
Industrial waste	0	0	0	0	2	2
Others	0	0	0	0	0	0
Total input	391344	368999	442213	408401	322976	317196
Total production	109606	103705	122886	117338	97276	96851
Estimated efficiency (% of production to input)	28	28	28	29	30	31

Statistics on electricity

Russian Federation

Item	2009	2010	2011	2012	2013	2014
Production, trade and consumption	**Million kilowatt-hours**					
Total main activity and autoproducer	991980	1038030	1054765	1070734	1059092	1064207
Combustible fuels	651810	698709	713689	725399	703481	705598
Hydro	176118	168397	167608	167319	182654	177141
Nuclear	163584	170415	172941	177534	172508	180757
Other	468	509	527	482	449	711
Main activity	939616	976126	990657	1003553	994597	997475
Combustible fuels	600378	637713	650687	659241	640434	640412
Hydro	175186	167489	166502	166296	181206	175595
Nuclear	163584	170415	172941	177534	172508	180757
Other	468	509	527	482	449	711
Own use in electricity, CHP and heat plants	59456	68522	68259	69232	64342	69550
Net production	932524	969508	986506	1001502	994750	994657
Imports	3066	1644	1558	2661	4706	6623
Exports	17923	19091	24111	19143	18382	14671
Losses	106792	104933	105002	106667	106994	106567
Consumption	810645	853369	858951	876939	874080	880042
Energy industries own use	124409	126686	130127	136654	129989	142212
By industry and construction	311417	326849	332818	338608	336692	333701
By transport	81206	85284	90355	92041	90559	90256
By households and other cons.	293613	314550	305651	309636	316840	313873
Net installed capacity	**Thousand kilowatts**					
Total main activity and autoproducer	225499	222935	222984	232685	239423	259020
Combustible fuels	153800	151114	151114	157845	163926	182465
Hydro	47308	47430	47479	49445	50104	50845
Nuclear	24300	24300	24300	25304	25304	25304
Other	91	91	91	91	89	406
Main activity	208283	205405	205454	214320	219789	233660
Combustible fuels	137143	134143	134143	139784	144744	157592
Hydro	46749	46871	46920	49141	49652	50358
Nuclear	24300	24300	24300	25304	25304	25304
Other	91	91	91	91	89	406
Combustible fuel input	**Terajoules**					
Hard Coal	1507270	1515713	1515618	1704144	1412388	1319217
Brown Coal	870435	944802	957586	947024	908342	857293
Peat	6178	6537	9178	6886	12098	11041
Gas-diesel oil	56502	59770	59942	59727	56373	58480
Fuel oil	176104	93809	324170	332977	82820	90658
Refinery gas	2277	2426	2673	2030	1931	2079
Natural gas	6791755	7399820	7876206	8031618	8427679	7755898
Coke-oven gas	63540	73905	74001	70030	70600	69246
Blast furnace gas	119057	122886	123099	42842	51219	52344
Industrial waste	69391	79777	80023	79347	80246	82455
Others	1236	1297	1348	1437	1213	1152
Total input	9663745	10300742	11023843	11278062	11104908	10299863
Total production	2346516	2515352	2569280	2611436	2532532	2540153
Estimated efficiency (% of production to input)	24	24	23	23	23	25

Statistics on electricity

Rwanda

Item	2009	2010	2011	2012	2013	2014
Production, trade and consumption	**Million kilowatt-hours**					
Total main activity and autoproducer	248	281	346	393	413	476
Combustible fuels	149	169	196	211	265	290
Hydro	99	112	149	182	148	186
Nuclear	..	..	..	..	..	..
Other	0	0	0	0	0	0
Main activity	248	281	346	393	413	476
Combustible fuels	149	169	196	211	265	290
Hydro	99	112	149	182	148	186
Nuclear	..	..	..	..	..	..
Other	0	0	0	0	0	0
Own use in electricity, CHP and heat plants	18	21	24	24	24	38
Net production	231	261	322	369	389	438
Imports	62	80	78	91	94	90
Exports	3	3	5	3	0	4
Losses	*45	*51	*68	*78	*95	*86
Consumption	246	287	326	380	388	438
Energy industries own use	..	..	..	..	..	..
By industry and construction	56	63	68	77	79	83
By transport	..	..	..	..	..	..
By households and other cons.	190	224	259	302	309	355
Net installed capacity	**Thousand kilowatts**					
Total main activity and autoproducer	75	84	89	*99	*109	*119
Combustible fuels	33	33	33	*36	*40	*42
Hydro	42	51	57	*63	*69	*76
Nuclear	..	..	..	..	..	..
Other	0	0	0	0	0	*1
Main activity	75	84	89	*99	*109	*119
Combustible fuels	33	33	33	*36	*40	*42
Hydro	42	51	57	*63	*69	*76
Nuclear	..	..	..	..	..	..
Other	0	0	0	0	0	*1
Combustible fuel input	**Terajoules**					
Gas-diesel oil	*1677	*1892	*2103	*2305	*2679	*2924
Total input	*1677	*1892	*2103	*2305	*2679	*2924
Total production	537	608	705	760	954	1044
Estimated efficiency (% of production to input)	32	32	34	33	36	36

Statistics on electricity

Samoa

Item	2009	2010	2011	2012	2013	2014
Production, trade and consumption	Million kilowatt-hours					
Total main activity and autoproducer	113	116	114	117	*121	*121
Combustible fuels	74	69	79	80	*85	*85
Hydro	39	48	35	37	*36	*36
Nuclear	..	..	..	..	..	..
Other	..	..	..	..	..	..
Main activity	108	111	109	112	*116	*116
Combustible fuels	69	64	74	75	*80	*80
Hydro	39	48	35	37	*36	*36
Nuclear	..	..	..	..	..	..
Other	..	..	..	..	..	..
Own use in electricity, CHP and heat plants	1	1	1	*1	*4	*4
Net production	112	116	114	117	*117	*117
Imports	..	..	..	..	..	..
Exports	..	..	..	..	..	..
Losses	*13	*16	*16	*16	*19	*19
Consumption	91	91	89	*93	*98	*98
Energy industries own use	..	..	..	..	..	..
By industry and construction	5	5	6	*6	*18	*18
By transport	..	..	..	..	..	..
By households and other cons.	87	86	83	*87	*80	*80
Net installed capacity	Thousand kilowatts					
Total main activity and autoproducer	42	*42	*42	*42	*42	38
Combustible fuels	30	*30	*30	*30	*30	26
Hydro	12	*12	*12	*12	*12	12
Nuclear	..	..	..	..	..	..
Other	..	..	..	..	..	..
Main activity	39	*39	*39	*39	*39	35
Combustible fuels	27	*27	*27	*27	*27	23
Hydro	12	*12	*12	*12	*12	12
Nuclear	..	..	..	..	..	..
Other	..	..	..	..	..	..
Combustible fuel input	Terajoules					
Gas-diesel oil	*757	731	808	*757	860	*860
Total input	*757	731	808	*757	860	*860
Total production	267	247	284	289	*306	*306
Estimated efficiency (% of production to input)	35	34	35	38	36	36

173

Statistics on electricity

Sao Tome and Principe

Item	2009	2010	2011	2012	2013	2014
Production, trade and consumption			Million kilowatt-hours			
Total main activity and autoproducer	*52	57	*60	*67	*67	*67
Combustible fuels	*45	52	*54	*60	*60	*60
Hydro	*7	5	*6	*7	*7	*7
Nuclear	..	..	..	..	..	..
Other	..	..	..	..	..	..
Main activity	*52	57	*60	*67	*67	*67
Combustible fuels	*45	52	*54	*60	*60	*60
Hydro	*7	5	*6	*7	*7	*7
Nuclear	..	..	..	..	..	..
Other	..	..	..	..	..	..
Own use in electricity, CHP and heat plants	0	0	0	0	0	0
Net production	*52	57	*60	*67	*67	*67
Imports	..	..	..	..	..	..
Exports	..	..	..	..	..	..
Losses	*15	15	*16	*16	*16	*16
Consumption	*38	42	*44	*51	*51	*51
Energy industries own use	..	..	..	..	..	..
By industry and construction	..	..	..	..	..	..
By transport	..	..	..	..	..	..
By households and other cons.	*38	42	*44	*51	*51	*51
Net installed capacity			Thousand kilowatts			
Total main activity and autoproducer	14	14	18	*18	*19	19
Combustible fuels	12	12	16	*16	*16	16
Hydro	2	2	*2	*2	*3	*3
Nuclear	..	..	..	..	..	..
Other	..	..	..	..	..	..
Main activity	14	14	18	*18	*19	19
Combustible fuels	12	12	16	*16	*16	16
Hydro	2	2	*2	*2	*3	*3
Nuclear	..	..	..	..	..	..
Other	..	..	..	..	..	..
Combustible fuel input			Terajoules			
Gas-diesel oil	*482	559	*581	*624	*624	*624
Total input	*482	559	*581	*624	*624	*624
Total production	*162	189	*194	*216	*216	*216
Estimated efficiency (% of production to input)	34	34	33	35	35	35

Statistics on electricity

Saudi Arabia

Item	2009	2010	2011	2012	2013	2014
Production, trade and consumption	**Million kilowatt-hours**					
Total main activity and autoproducer	217082	240067	250077	271680	284017	311806
Combustible fuels	217082	240067	250077	271679	284016	311805
Hydro	..	..	..	..	..	..
Nuclear	..	..	..	..	..	..
Other	0	0	0	1	1	1
Main activity	184070	186482	190280	207132	198900	214588
Combustible fuels	184070	186482	190280	207131	198899	214587
Hydro	..	..	..	..	..	..
Nuclear	..	..	..	..	..	..
Other	0	0	0	1	1	1
Own use in electricity, CHP and heat plants	5645	5696	6128	6826	6562	7566
Net production	211437	234371	243949	264854	277455	304240
Imports	..	..	..	..	..	..
Exports	..	..	..	..	..	..
Losses	17965	21388	23508	23780	20017	21144
Consumption	193472	212263	219662	241074	257438	283096
Energy industries own use	9591	9444	9269	9767	9212	10981
By industry and construction	24682	28619	32320	32075	41947	48442
By transport	..	..	..	..	..	..
By households and other cons.	159199	174200	178073	199232	206279	223673
Net installed capacity	**Thousand kilowatts**					
Total main activity and autoproducer	44582	46374	51149	62133	*58472	*65507
Combustible fuels	44582	46374	51149	62132	*58471	*65506
Hydro	..	..	..	..	..	..
Nuclear	..	..	..	..	..	..
Other	..	0	0	*1	*1	*1
Main activity	37802	36023	38919	47276	*44432	*49778
Combustible fuels	37802	36023	38919	47275	*44431	*49777
Hydro	..	..	..	..	..	..
Nuclear	..	..	..	..	..	..
Other	..	0	0	*1	*1	*1
Combustible fuel input	**Terajoules**					
Crude oil	713897	857506	926455	920575	812710	952384
Gas-diesel oil	477085	486158	547304	575383	592540	541155
Fuel oil	161600	103505	148026	178164	181234	262398
Natural gas	1288608	1403473	1380378	1538835	1751556	1801750
Total input	2641190	2850641	3002162	3212957	3338040	3557688
Total production	781495	864241	900277	978044	1022458	1122498
Estimated efficiency (% of production to input)	30	30	30	30	31	32

175

Statistics on electricity

Senegal

Item	2009	2010	2011	2012	2013	2014
Production, trade and consumption	**Million kilowatt-hours**					
Total main activity and autoproducer	2489	2618	2560	2917	3403	3406
Combustible fuels	2489	2618	2560	2917	3335	3335
Hydro	..	..	..	..	..	..
Nuclear	..	..	..	..	..	..
Other	..	..	..	..	68	71
Main activity	1895	1800	1357	1421	2730	2858
Combustible fuels	1895	1800	1357	1421	2730	2858
Hydro	..	..	..	..	..	..
Nuclear	..	..	..	..	..	..
Other	..	..	..	..	..	..
Own use in electricity, CHP and heat plants	62	58	34	41	39	36
Net production	2427	2560	2526	2876	3364	3370
Imports	239	253	257	290	308	318
Exports	..	..	..	..	..	..
Losses	*560	269	265	302	593	478
Consumption	2361	2057	2032	2313	3079	3210
Energy industries own use	..	..	..	..	..	..
By industry and construction	*663	738	751	858	846	887
By transport	..	..	..	..	..	..
By households and other cons.	1698	1319	1281	1455	2233	2323
Net installed capacity	**Thousand kilowatts**					
Total main activity and autoproducer	*639	686	827	854	*854	*875
Combustible fuels	*636	686	827	854	*854	*854
Hydro	..	..	..	..	..	..
Nuclear	..	..	..	..	..	..
Other	*3	..	..	..	..	*21
Main activity	*477	493	483	489	*489	*489
Combustible fuels	*477	493	483	489	*489	*489
Hydro	..	..	..	..	..	..
Nuclear	..	..	..	..	..	..
Other	..	..	..	..	..	..
Combustible fuel input	**Terajoules**					
Gas-diesel oil	*1763	5117	7224	8686	5891	6192
Fuel oil	*16362	21533	18099	18099	21735	22745
Natural gas	677	971	752	1083	1519	1591
Bagasse	*991	*2809	*2831	*2229	*1370	*1435
Total input	*19793	30430	28906	30097	30515	31963
Total production	8960	9425	9216	10501	12006	12006
Estimated efficiency (% of production to input)	45	31	32	35	39	38

Statistics on electricity

Serbia

Item	2009	2010	2011	2012	2013	2014
Production, trade and consumption	Million kilowatt-hours					
Total main activity and autoproducer	38322	38103	38600	36799	39877	34060
Combustible fuels	27178	25532	29357	26885	29024	22437
Hydro	11144	12571	9243	9914	10853	11617
Nuclear	..	..	..	..	..	..
Other	..	..	..	..	..	6
Main activity	38162	37817	38370	36628	39675	33771
Combustible fuels	27018	25246	29127	26714	28822	22148
Hydro	11144	12571	9243	9914	10853	11617
Nuclear	..	..	..	..	..	..
Other	..	..	..	..	..	6
Own use in electricity, CHP and heat plants	2553	2402	2807	2597	2707	2161
Net production	35769	35701	35793	34202	37170	31899
Imports	5184	5620	6701	5781	4077	7008
Exports	6609	5917	6979	5392	6614	5445
Losses	5971	6024	5844	5609	5501	5163
Consumption	28373	29380	29671	28982	29132	28299
Energy industries own use	1563	1811	1680	1815	2229	2141
By industry and construction	6772	7287	7473	6931	7079	7156
By transport	502	224	529	492	478	336
By households and other cons.	19536	20058	19989	19744	19346	18666
Net installed capacity	Thousand kilowatts					
Total main activity and autoproducer	8385	*8714	*8714	*8714	*8714	8385
Combustible fuels	5554	*5559	*5559	*5559	*5559	5175
Hydro	2831	*2835	*2835	*2835	*2835	2835
Nuclear	..	..	..	..	..	..
Other	..	320	320	320	*320	375
Main activity	8355	*8679	*8679	*8679	*8679	8350
Combustible fuels	5524	*5524	*5524	*5524	*5524	5140
Hydro	2831	*2835	*2835	*2835	*2835	2835
Nuclear	..	..	..	..	..	..
Other	..	320	320	320	*320	375
Combustible fuel input	Terajoules					
Brown Coal	292961	280451	316532	290771	304399	233364
Lignite briquettes	105	896	860	736	667	389
Biogas	..	..	..	39	171	126
Gas-diesel oil	0	0	0	0	0	0
Fuel oil	1778	5252	4969	2990	485	323
Liquefied petroleum gas	0	0	0	0	0	0
Natural gas	4573	8846	10298	8588	8033	7087
Blast furnace gas	1714	3753	3757	991	1142	1841
Fuelwood	0	0	0	0	78	78
Industrial waste	0	0	0	50	38	34
Others	0	69	0	0	140	0
Total input	301131	299268	336416	304164	315153	243242
Total production	97841	91915	105685	96786	104486	80773
Estimated efficiency (% of production to input)	32	31	31	32	33	33

Seychelles

Item	2009	2010	2011	2012	2013	2014
Production, trade and consumption	**Million kilowatt-hours**					
Total main activity and autoproducer	276	301	324	336	354	363
Combustible fuels	276	301	324	336	347	355
Hydro	..	..	..	..	..	..
Nuclear	..	..	..	..	..	..
Other	..	..	..	..	7	8
Main activity	276	301	324	336	354	363
Combustible fuels	276	301	324	336	347	355
Hydro	..	..	..	..	..	..
Nuclear	..	..	..	..	..	..
Other	..	..	..	..	7	8
Own use in electricity, CHP and heat plants	10	11	10	9	10	10
Net production	266	290	314	327	344	352
Imports	..	..	..	..	..	..
Exports	..	..	..	..	..	..
Losses	32	36	37	40	40	41
Consumption	239	260	280	314	307	313
Energy industries own use	..	..	..	..	..	..
By industry and construction	106	118	138	165	177	182
By transport	..	..	..	..	..	..
By households and other cons.	133	142	142	148	130	131
Net installed capacity	**Thousand kilowatts**					
Total main activity and autoproducer	63	63	79	79	85	86
Combustible fuels	63	63	79	79	79	80
Hydro	..	..	..	..	..	..
Nuclear	..	..	..	..	..	..
Other	..	0	0	0	6	6
Main activity	63	63	79	79	85	86
Combustible fuels	63	63	79	79	79	80
Hydro	..	..	..	..	..	..
Nuclear	..	..	..	..	..	..
Other	..	0	0	0	6	6
Combustible fuel input	**Terajoules**					
Gas-diesel oil	110	318	121	30	15	*15
Fuel oil	2065	2098	2467	2640	2705	*2945
Total input	2175	2416	2588	2670	2720	*2960
Total production	992	1084	1166	1211	1250	1278
Estimated efficiency (% of production to input)	46	45	45	45	46	43

Statistics on electricity

Sierra Leone

Item	2009	2010	2011	2012	2013	2014
Production, trade and consumption	**Million kilowatt-hours**					
Total main activity and autoproducer	*132	*171	176	179	164	181
Combustible fuels	44	22	22	61	59	63
Hydro	*88	*149	153	118	105	118
Nuclear	..	..	..	..	..	..
Other	..	..	..	..	..	..
Main activity	*132	*171	176	179	164	181
Combustible fuels	44	22	22	61	59	63
Hydro	*88	*149	153	118	105	118
Nuclear	..	..	..	..	..	..
Other	..	..	..	..	..	..
Own use in electricity, CHP and heat plants	1	3	1	0	0	2
Net production	*131	*167	175	178	164	178
Imports	..	..	..	..	..	..
Exports	..	..	..	..	..	..
Losses	64	82	89	78	95	78
Consumption	68	85	86	100	69	86
Energy industries own use	..	..	..	..	..	..
By industry and construction	23	25	35	32	29	35
By transport	..	..	..	..	..	..
By households and other cons.	46	60	52	67	40	52
Net installed capacity	**Thousand kilowatts**					
Total main activity and autoproducer	*77	77	77	77	77	77
Combustible fuels	27	27	27	27	27	27
Hydro	*50	50	50	50	50	50
Nuclear	..	..	..	..	..	..
Other	..	..	..	..	..	..
Main activity	*77	77	77	77	77	77
Combustible fuels	27	27	27	27	27	27
Hydro	*50	50	50	50	50	50
Nuclear	..	..	..	..	..	..
Other	..	..	..	..	..	..
Combustible fuel input	**Terajoules**					
Gas-diesel oil	*495	*258	*258	*688	*774	*817
Total input	*495	*258	*258	*688	*774	*817
Total production	158	78	80	218	212	225
Estimated efficiency (% of production to input)	32	30	31	32	27	28

Statistics on electricity

Singapore

Item	2009	2010	2011	2012	2013	2014
Production, trade and consumption	**Million kilowatt-hours**					
Total main activity and autoproducer	41815	45361	46000	46971	47964	49310
Combustible fuels	41815	45361	46000	46971	47964	49310
Hydro	..	..	..	..	..	..
Nuclear	..	..	..	..	..	..
Other	..	..	..	..	..	..
Main activity	40671	44097	44685	45197	45160	46336
Combustible fuels	40671	44097	44685	45197	45160	46336
Hydro	..	..	..	..	..	..
Nuclear	..	..	..	..	..	..
Other	..	..	..	..	..	..
Own use in electricity, CHP and heat plants	1339	1405	1358	1310	2007	2000
Net production	40476	43956	44642	45661	45957	47310
Imports	..	..	..	..	..	..
Exports	..	..	..	..	..	..
Losses	1091	*1162	*1115	*1034	*1034	*1035
Consumption	38836	42266	42821	44315	44963	46417
Energy industries own use	0	0	0	0	0	0
By industry and construction	*15570	*17663	*17877	*18672	*18843	*19753
By transport	1608	2099	2224	2329	2369	2441
By households and other cons.	21658	22505	22720	23314	23751	24223
Net installed capacity	**Thousand kilowatts**					
Total main activity and autoproducer	10337	10225	10131	10296	11425	13047
Combustible fuels	10337	10225	10131	10296	11425	13047
Hydro	..	..	..	..	..	..
Nuclear	..	..	..	..	..	..
Other	..	..	..	..	..	..
Main activity	10017	9917	9825	9982	10923	12545
Combustible fuels	10017	9917	9825	9982	10923	12545
Hydro	..	..	..	..	..	..
Nuclear	..	..	..	..	..	..
Other	..	..	..	..	..	..
Combustible fuel input	**Terajoules**					
Hard Coal	..	..	..	79	5631	9598
Gas-diesel oil	774	5977	9675	5762	1075	86
Fuel oil	60075	73649	69407	49732	19917	2586
Refinery gas	13464	13514	14256	12227	10841	8366
Natural gas	265462	277515	285758	314224	353928	375435
Municipal waste	18372	24589	25263	26008	26917	27063
Total input	358146	395244	404360	408033	418309	423133
Total production	150532	163298	165601	169097	172672	177516
Estimated efficiency (% of production to input)	42	41	41	41	41	42

Statistics on electricity

Sint Maarten (Dutch part)

Item	2009	2010	2011	2012	2013	2014
Production, trade and consumption	Million kilowatt-hours					
Total main activity and autoproducer	..	..	..	*402	410	*418
Combustible fuels	..	..	..	*402	410	*418
Hydro	..	..	..	..	..	..
Nuclear	..	..	..	..	..	..
Other	..	..	..	..	..	..
Main activity	..	..	..	*402	410	*418
Combustible fuels	..	..	..	*402	410	*418
Hydro	..	..	..	..	..	..
Nuclear	..	..	..	..	..	..
Other	..	..	..	..	..	..
Own use in electricity, CHP and heat plants	..	..	..	0	0	0
Net production	..	..	..	*402	410	*418
Imports	..	..	..	..	..	..
Exports	..	..	..	..	..	..
Losses	..	..	..	*36	37	*38
Consumption	..	..	..	*366	373	*380
Energy industries own use	..	..	..	..	..	..
By industry and construction	..	..	..	*61	62	*63
By transport	..	..	..	..	..	..
By households and other cons.	..	..	..	*305	311	*317
Net installed capacity	Thousand kilowatts					
Total main activity and autoproducer	..	..	..	97	97	97
Combustible fuels	..	..	..	97	97	97
Hydro	..	..	..	..	..	..
Nuclear	..	..	..	..	..	..
Other	..	..	..	..	..	..
Main activity	..	..	..	97	97	97
Combustible fuels	..	..	..	97	97	97
Hydro	..	..	..	..	..	..
Nuclear	..	..	..	..	..	..
Other	..	..	..	..	..	..
Combustible fuel input	Terajoules					
Fuel oil	..	..	..	*3091	3167	*3248
Total input	..	..	..	*3091	3167	*3248
Total production	..	..	..	*1446	1475	*1504
Estimated efficiency (% of production to input)	..	..	..	47	47	46

181

Statistics on electricity

Slovakia

Item	2009	2010	2011	2012	2013	2014
Production, trade and consumption	**Million kilowatt-hours**					
Total main activity and autoproducer	26155	27858	28656	28664	28832	27401
Combustible fuels	7425	7572	8619	8231	7283	6719
Hydro	4604	5649	4146	4439	5166	4462
Nuclear	14081	14574	15411	15495	15720	15499
Other	45	63	480	499	663	721
Main activity	24065	25818	26164	26128	26362	24361
Combustible fuels	5488	5703	6623	6192	5458	4402
Hydro	4496	5539	4045	4351	5060	4346
Nuclear	14081	14574	15411	15495	15720	15499
Other	..	2	85	90	124	114
Own use in electricity, CHP and heat plants	2053	2427	2566	2527	1660	2394
Net production	24102	25431	26090	26137	27172	25007
Imports	8994	7334	11227	13472	10719	12964
Exports	7682	6293	10500	13079	10628	11862
Losses	782	856	515	1274	760	667
Consumption	24629	25613	26299	25237	26491	25426
Energy industries own use	1531	1478	1487	1300	1407	1269
By industry and construction	10768	10927	11241	11908	11785	12237
By transport	503	538	536	560	567	574
By households and other cons.	11827	12670	13035	11469	12732	11346
Net installed capacity	**Thousand kilowatts**					
Total main activity and autoproducer	7153	7873	8363	8412	8458	8092
Combustible fuels	2824	3496	3382	3416	3432	3068
Hydro	2487	2516	2523	2522	2523	2523
Nuclear	1820	1820	1940	1939	1940	1940
Other	22	41	518	535	563	561
Main activity	6528	7258	7362	7370	7385	6947
Combustible fuels	2246	2945	2820	2833	2847	2411
Hydro	2462	2491	2494	2493	2493	2493
Nuclear	1820	1820	1940	1939	1940	1940
Other	..	2	108	105	105	103
Combustible fuel input	**Terajoules**					
Hard Coal	21706	20061	18980	18850	17031	15816
Brown Coal	32436	33174	34024	30403	28653	27261
Biogas	268	345	1072	1776	2072	3857
Fuel oil	10544	12201	11150	10262	8928	6504
Natural gas	22067	26043	30484	28153	26454	19005
Coke-oven gas	1021	1109	1102	1095	887	961
Blast furnace gas	1459	1818	1756	1857	1697	1661
Fuelwood	*2701	*4431	*4946	*8487	*7807	*5259
Vegetal waste	*683	*1355	*1333	*425	*785	*1035
Black liquor	*2667	*4041	*5106	*7630	*8053	*7169
Others	1678	1762	1606	1785	2485	2408
Total input	97230	106340	111559	110723	104852	90936
Total production	26730	27259	31028	29632	26219	24188
Estimated efficiency (% of production to input)	27	26	28	27	25	27

Slovenia

Item	2009	2010	2011	2012	2013	2014
Production, trade and consumption	Million kilowatt-hours					
Total main activity and autoproducer	16403	16440	16059	15736	16103	17437
Combustible fuels	5945	6067	6073	5958	5661	4440
Hydro	4715	4703	3706	4087	4923	6366
Nuclear	5739	5657	6215	5528	5300	6370
Other	4	13	65	163	219	261
Main activity	15935	15969	15623	15202	15481	16688
Combustible fuels	5645	5774	5825	5707	5414	4169
Hydro	4550	4537	3581	3961	4757	6138
Nuclear	5739	5657	6215	5528	5300	6370
Other	1	1	2	6	10	11
Own use in electricity, CHP and heat plants	1028	1030	1058	1031	986	951
Net production	15375	15410	15001	14705	15117	16486
Imports	7780	8625	7036	7452	7521	7254
Exports	10839	10745	8408	8491	8811	9997
Losses	893	982	824	875	848	821
Consumption	11423	12308	12805	12791	12979	12922
Energy industries own use	130	363	305	363	500	463
By industry and construction	4966	5487	5864	5922	5878	6057
By transport	156	173	164	159	154	136
By households and other cons.	6171	6285	6472	6347	6447	6266
Net installed capacity	Thousand kilowatts					
Total main activity and autoproducer	3050	3193	3268	3351	3434	3454
Combustible fuels	1310	1261	1270	1267	1256	1242
Hydro	1070	1254	1253	1254	1299	1296
Nuclear	666	666	688	688	688	688
Other	4	12	57	142	191	228
Main activity	2863	3009	3040	3058	3095	3070
Combustible fuels	1242	1204	1213	1227	1214	1192
Hydro	954	1138	1137	1138	1183	1180
Nuclear	666	666	688	688	688	688
Other	1	1	2	5	10	10
Combustible fuel input	Terajoules					
Hard Coal	48	101	51	76	131	101
Brown Coal	57005	57229	58793	56744	53877	42032
Biogas	850	1166	1404	1497	1357	1203
Gas-diesel oil	215	86	129	129	86	344
Fuel oil	81	40	40	0	0	0
Natural gas	5815	5213	4789	5158	4824	3153
Fuelwood	*1619	*1524	*1566	*1380	*1495	*1467
Industrial waste	24	21	263	316	310	321
Biodiesel	..	..	..	..	..	37
Total input	65656	65380	67035	65300	62080	48658
Total production	21402	21841	21863	21449	20380	15984
Estimated efficiency (% of production to input)	33	33	33	33	33	33

Solomon Islands

Item	2009	2010	2011	2012	2013	2014
Production, trade and consumption	**Million kilowatt-hours**					
Total main activity and autoproducer	84	87	83	83	89	92
Combustible fuels	84	87	83	83	89	92
Hydro	..	..	..	..	..	..
Nuclear	..	..	..	..	..	..
Other	..	..	..	..	..	..
Main activity	76	79	75	75	81	84
Combustible fuels	76	79	75	75	81	84
Hydro	..	..	..	..	..	..
Nuclear	..	..	..	..	..	..
Other	..	..	..	..	..	..
Own use in electricity, CHP and heat plants	*2	*2	2	2	*2	*2
Net production	82	85	81	81	87	90
Imports	..	..	..	..	..	..
Exports	..	..	..	..	..	..
Losses	*10	*11	*10	*10	*10	*10
Consumption	*72	*75	71	*71	77	*80
Energy industries own use	..	..	..	..	..	..
By industry and construction	*27	*29	*31	*31	*30	*35
By transport	..	..	..	..	..	..
By households and other cons.	*45	*46	40	*41	47	*45
Net installed capacity	**Thousand kilowatts**					
Total main activity and autoproducer	*36	*36	*36	*36	*36	*36
Combustible fuels	*36	*36	*36	*36	*36	*36
Hydro	0	0	0	0	0	0
Nuclear	..	..	..	..	..	..
Other	..	..	..	..	..	..
Main activity	*31	*31	*31	*31	*31	*31
Combustible fuels	*31	*31	*31	*31	*31	*31
Hydro	0	0	0	0	0	0
Nuclear	..	..	..	..	..	..
Other	..	..	..	..	..	..
Combustible fuel input	**Terajoules**					
Gas-diesel oil	*1221	*1299	*1226	*1234	*1277	*1294
Total input	*1221	*1299	*1226	*1234	*1277	*1294
Total production	301	315	298	300	321	331
Estimated efficiency (% of production to input)	25	24	24	24	25	26

Statistics on electricity

Somalia

Item	2009	2010	2011	2012	2013	2014
Production, trade and consumption	Million kilowatt-hours					
Total main activity and autoproducer	327	327	334	340	350	350
Combustible fuels	327	327	334	340	350	350
Hydro	..	..	..	..	..	..
Nuclear	..	..	..	..	..	..
Other	..	..	..	..	..	..
Main activity	327	327	334	340	350	350
Combustible fuels	327	327	334	340	350	350
Hydro	..	..	..	..	..	..
Nuclear	..	..	..	..	..	..
Other	..	..	..	..	..	..
Own use in electricity, CHP and heat plants	0	0	0	0	0	0
Net production	327	327	334	340	350	350
Imports	..	..	..	..	..	..
Exports	..	..	..	..	..	..
Losses	*32	*32	*33	*32	*35	*35
Consumption	292	295	301	308	315	315
Energy industries own use	..	..	..	..	..	..
By industry and construction	..	..	..	..	..	..
By transport	..	..	..	..	..	..
By households and other cons.	292	295	301	308	315	315
Net installed capacity	Thousand kilowatts					
Total main activity and autoproducer	80	80	80	80	80	80
Combustible fuels	80	80	80	80	80	80
Hydro	..	..	..	..	..	..
Nuclear	..	..	..	..	..	..
Other	..	..	..	..	..	..
Main activity	80	80	80	80	80	80
Combustible fuels	80	80	80	80	80	80
Hydro	..	..	..	..	..	..
Nuclear	..	..	..	..	..	..
Other	..	..	..	..	..	..
Combustible fuel input	Terajoules					
Gas-diesel oil	*3612	*3612	*3655	*3698	*3698	*3698
Total input	*3612	*3612	*3655	*3698	*3698	*3698
Total production	1177	1177	1202	1224	1260	1260
Estimated efficiency (% of production to input)	33	33	33	33	34	34

Statistics on electricity

South Africa

Item	2009	2010	2011	2012	2013	2014
Production, trade and consumption	**Million kilowatt-hours**					
Total main activity and autoproducer	249557	259608	262538	257919	256073	252578
Combustible fuels	232503	242401	243979	241678	237647	232512
Hydro	4142	5067	5019	4211	4040	4082
Nuclear	12806	12099	13502	11954	14106	13794
Other	106	41	38	76	280	2190
Main activity	241093	251264	251746	247516	244851	240401
Combustible fuels	224165	234211	233340	231403	226548	220459
Hydro	4016	4913	4866	4083	3917	3958
Nuclear	12806	12099	13502	11954	14106	13794
Other	106	41	38	76	280	2190
Own use in electricity, CHP and heat plants	14510	14896	14963	14683	14605	14306
Net production	235047	244712	247575	243236	241468	238272
Imports	12295	12193	11890	10006	9428	11177
Exports	14052	14668	14964	15035	13929	13836
Losses	24280	24468	21998	22221	21488	20940
Consumption	209010	217769	222503	215986	215479	214673
Energy industries own use	14589	15128	15262	15873	16171	16580
By industry and construction	114047	120081	124014	121354	120959	120260
By transport	3539	3591	3772	3818	3798	3773
By households and other cons.	76835	78969	79455	74941	74551	74060
Net installed capacity	**Thousand kilowatts**					
Total main activity and autoproducer	*44813	*44813	*44813	*44820	*44170	*46963
Combustible fuels	*42186	*42186	*42186	*42186	*41559	*43538
Hydro	*714	*714	*714	*714	*714	*725
Nuclear	*1890	*1890	*1890	*1890	*1800	*1880
Other	*23	*23	*23	*30	*97	*820
Main activity	*42777	*42777	*42777	*42784	*42134	*44823
Combustible fuels	*40186	*40186	*40186	*40186	*39559	*41438
Hydro	*678	*678	*678	*678	*678	*685
Nuclear	*1890	*1890	*1890	*1890	*1800	*1880
Other	*23	*23	*23	*30	*97	*820
Combustible fuel input	**Terajoules**					
Hard Coal	2813736	2994012	2840439	2915759	2938789	3103299
Gas-diesel oil	516	2021	2021	1978	1978	1935
Vegetal waste	*4046	*4104	*4162	*4219	*4291	*4363
Total input	2818298	3000137	2846622	2921956	2945058	3109597
Total production	837011	872644	878324	870041	855529	837043
Estimated efficiency (% of production to input)	30	29	31	30	29	27

South Sudan

Item	2009	2010	2011	2012	2013	2014
Production, trade and consumption				**Million kilowatt-hours**		
Total main activity and autoproducer	..	..	..	445	472	488
Combustible fuels	..	..	..	443	470	486
Hydro	..	..	..	..	..	..
Nuclear	..	..	..	..	..	..
Other	..	..	..	2	2	2
Main activity	..	..	..	211	226	234
Combustible fuels	..	..	..	210	225	233
Hydro	..	..	..	..	..	..
Nuclear	..	..	..	..	..	..
Other	..	..	..	1	1	1
Own use in electricity, CHP and heat plants	..	..	..	18	18	19
Net production	..	..	..	427	454	469
Imports	..	..	..	..	..	..
Exports	..	..	..	..	..	..
Losses	..	..	..	26	27	28
Consumption	..	..	..	385	415	429
Energy industries own use	..	..	..	..	..	..
By industry and construction	..	..	..	..	..	..
By transport	..	..	..	..	..	..
By households and other cons.	..	..	..	385	415	429
Net installed capacity				**Thousand kilowatts**		
Total main activity and autoproducer	..	..	..	*82	*82	*82
Combustible fuels	..	..	..	*80	*80	*80
Hydro	..	..	..	..	..	..
Nuclear	..	..	..	..	..	..
Other	..	..	..	*2	*2	*2
Main activity	..	..	..	*41	*41	*41
Combustible fuels	..	..	..	*40	*40	*40
Hydro	..	..	..	..	..	..
Nuclear	..	..	..	..	..	..
Other	..	..	..	*1	*1	*1
Combustible fuel input				**Terajoules**		
Gas-diesel oil	..	..	..	5074	5418	5590
Total input	..	..	..	5074	5418	5590
Total production	..	..	..	1595	1692	1750
Estimated efficiency (% of production to input)	..	..	..	31	31	31

Statistics on electricity

Spain

Item	2009	2010	2011	2012	2013	2014
Production, trade and consumption	**Million kilowatt-hours**					
Total main activity and autoproducer	294620	301527	293848	297559	285632	278750
Combustible fuels	168175	142410	150730	150312	119111	112789
Hydro	29162	45511	32911	24162	41052	42970
Nuclear	52761	61990	57718	61470	56726	57305
Other	44522	51616	52489	61615	68743	65686
Main activity	256495	262758	253030	256132	247504	246639
Combustible fuels	131038	104688	110980	109560	81966	81783
Hydro	28553	44664	32050	23700	40109	41906
Nuclear	52761	61990	57718	61470	56726	57305
Other	44143	51416	52282	61402	68703	65645
Own use in electricity, CHP and heat plants	11173	10576	10570	10987	10234	10370
Net production	283447	290951	283278	286572	275398	268380
Imports	6751	5206	7932	7787	9887	12310
Exports	14855	13539	14023	18986	16638	15716
Losses	24445	27400	26027	25675	26694	26393
Consumption	250897	254627	252050	250526	241089	238508
Energy industries own use	11119	9825	8541	10278	11002	11611
By industry and construction	76796	73490	73451	72466	70828	71657
By transport	2985	3222	4515	4455	4552	4159
By households and other cons.	159997	168090	165543	163327	154707	151081
Net installed capacity	**Thousand kilowatts**					
Total main activity and autoproducer	103880	109092	110110	112475	113302	113774
Combustible fuels	54760	57457	56786	56736	56786	56786
Hydro	18705	18735	18740	18750	19385	19423
Nuclear	7365	7450	7450	7450	6984	7399
Other	23050	25450	27134	29539	30147	30166
Main activity	96576	101788	102806	105171	105998	106470
Combustible fuels	47760	50457	49786	49736	49786	49786
Hydro	18505	18535	18540	18550	19185	19223
Nuclear	7365	7450	7450	7450	6984	7399
Other	22946	25346	27030	29435	30043	30062
Combustible fuel input	**Terajoules**					
Hard Coal	319960	228811	412497	523188	379601	411565
Brown Coal	14042	8026	37963	30335	21620	29263
Biogas	6754	8562	8287	8946	10237	8737
Gas-diesel oil	48289	48590	47902	44720	40377	49106
Fuel oil	87789	84274	74134	78740	66579	63913
Petroleum coke	23205	10205	13780	16673	9458	10725
Natural gas	757935	680182	609377	517885	395682	323797
Blast furnace gas	7465	7617	9254	7243	9939	12500
Fuelwood	*18085	*20078	*24039	*26714	*25095	*25095
Municipal waste	26728	14586	16332	14710	16724	17098
Others	14739	*12448	*7243	*8408	16814	18988
Total input	1324992	1123379	1260809	1277561	992125	970787
Total production	605430	512676	542628	541123	428800	406040
Estimated efficiency (% of production to input)	46	46	43	42	43	42

Statistics on electricity

Sri Lanka

Item	2009	2010	2011	2012	2013	2014
Production, trade and consumption	Million kilowatt-hours					
Total main activity and autoproducer	9987	10801	11600	11897	12044	12832
Combustible fuels	6086	5096	6916	8439	4870	7986
Hydro	3891	5645	4584	3302	6929	4563
Nuclear	..	..	..	..	..	..
Other	10	60	100	156	245	283
Main activity	9970	10783	11582	11879	12024	12812
Combustible fuels	6086	5096	6916	8439	4870	7986
Hydro	3881	5634	4573	3291	6918	4552
Nuclear	..	..	..	..	..	..
Other	4	53	93	149	236	274
Own use in electricity, CHP and heat plants	167	134	274	296	264	408
Net production	9820	10667	11326	11601	11780	12424
Imports	..	..	..	..	..	..
Exports	..	..	..	..	..	..
Losses	1431	1458	1337	1192	1201	1426
Consumption	8389	9209	9990	10409	10554	10997
Energy industries own use	..	..	..	..	..	..
By industry and construction	2773	3148	3379	3528	3590	3758
By transport	..	..	..	..	..	..
By households and other cons.	5616	6060	6611	6881	6964	7239
Net installed capacity	Thousand kilowatts					
Total main activity and autoproducer	2708	2833	3155	3383	3309	4058
Combustible fuels	1327	1412	1712	1716	1601	2243
Hydro	1378	1383	1401	1585	1617	1664
Nuclear	..	..	..	..	..	..
Other	3	38	42	82	91	150
Main activity	2698	2818	3140	3368	3294	4043
Combustible fuels	1317	1402	1702	1706	1591	2233
Hydro	1378	1383	1401	1585	1617	1664
Nuclear	..	..	..	..	..	..
Other	3	33	37	77	86	145
Combustible fuel input	Terajoules					
Hard Coal	0	0	10419	18318	19860	39964
Gas-diesel oil	13158	8127	13072	16899	3856	12995
Fuel oil	33572	30219	30906	33532	21257	24659
Naphtha	4940	2403	2047	2759	4652	5904
Fuelwood	724	1024	997	698	..	..
Bagasse	..	..	..	..	334	524
Total input	52394	41773	57440	72205	49959	84047
Total production	21908	18345	24896	30379	17530	28748
Estimated efficiency (% of production to input)	42	44	43	42	35	34

189

St. Helena and Depend.

Item	2009	2010	2011	2012	2013	2014
Production, trade and consumption	**Million kilowatt-hours**					
Total main activity and autoproducer	9	9	9	10	11	*11
Combustible fuels	8	8	8	9	10	*10
Hydro	..	..	..	..	..	..
Nuclear	..	..	..	..	..	..
Other	1	1	1	1	1	*1
Main activity	9	9	9	10	11	*11
Combustible fuels	8	8	8	9	10	*10
Hydro	..	..	..	..	..	..
Nuclear	..	..	..	..	..	..
Other	1	1	1	1	1	*1
Own use in electricity, CHP and heat plants	0	1	1	0	1	*1
Net production	9	9	9	9	10	*10
Imports	..	..	..	..	..	..
Exports	..	..	..	..	..	..
Losses	0	*1	*1	*1	*1	*1
Consumption	8	8	8	9	9	*9
Energy industries own use	..	..	..	..	..	..
By industry and construction	..	..	..	..	..	..
By transport	..	..	..	..	..	..
By households and other cons.	8	8	8	9	9	*9
Net installed capacity	**Thousand kilowatts**					
Total main activity and autoproducer	5	5	5	8	8	9
Combustible fuels	5	5	5	8	8	8
Hydro	..	..	..	..	..	..
Nuclear	..	..	..	..	..	..
Other	0	0	0	1	1	1
Main activity	5	5	5	8	8	9
Combustible fuels	5	5	5	8	8	8
Hydro	..	..	..	..	..	..
Nuclear	..	..	..	..	..	..
Other	0	0	0	1	1	1
Combustible fuel input	**Terajoules**					
Gas-diesel oil	*89	*97	*99	*103	*112	*112
Total input	*89	*97	*99	*103	*112	*112
Total production	30	29	30	32	35	*35
Estimated efficiency (% of production to input)	34	30	30	31	31	31

Statistics on electricity

St. Kitts-Nevis

Item	2009	2010	2011	2012	2013	2014
Production, trade and consumption	**Million kilowatt-hours**					
Total main activity and autoproducer	*217	216	223	218	*220	*220
Combustible fuels	*217	211	217	210	*211	*211
Hydro	..	..	..	..	..	..
Nuclear	..	..	..	..	..	..
Other	0	6	6	8	*9	*9
Main activity	*216	216	222	217	*219	*219
Combustible fuels	*216	210	216	209	*210	*210
Hydro	..	..	..	..	..	..
Nuclear	..	..	..	..	..	..
Other	0	6	6	8	*9	*9
Own use in electricity, CHP and heat plants	*11	0	0	0	*1	*1
Net production	*206	216	222	218	*219	*219
Imports	..	..	..	..	..	..
Exports	..	..	..	..	..	..
Losses	*38	39	40	39	*39	*39
Consumption	*168	178	183	179	*180	*180
Energy industries own use	..	..	..	..	..	..
By industry and construction	*33	36	*37	*36	*36	*36
By transport	..	..	..	..	..	..
By households and other cons.	*135	142	146	143	*144	*144
Net installed capacity	**Thousand kilowatts**					
Total main activity and autoproducer	*55	*57	*57	*62	*62	*63
Combustible fuels	*55	*55	*55	*59	*59	*60
Hydro	..	..	..	..	..	..
Nuclear	..	..	..	..	..	..
Other	0	2	2	2	3	*3
Main activity	*51	*53	*53	*58	*58	*59
Combustible fuels	*51	*51	*51	*55	*55	*56
Hydro	..	..	..	..	..	..
Nuclear	..	..	..	..	..	..
Other	0	2	2	2	3	*3
Combustible fuel input	**Terajoules**					
Gas-diesel oil	*2068	2034	2116	2017	*1982	*1978
Total input	*2068	2034	2116	2017	*1982	*1978
Total production	*781	759	781	756	*760	*760
Estimated efficiency (% of production to input)	38	37	37	37	38	38

191

Statistics on electricity

St. Lucia

Item	2009	2010	2011	2012	2013	2014
Production, trade and consumption	**Million kilowatt-hours**					
Total main activity and autoproducer	363	381	385	385	383	379
Combustible fuels	363	381	385	385	383	379
Hydro	..	..	..	..	..	..
Nuclear	..	..	..	..	..	..
Other	..	..	..	..	..	..
Main activity	363	381	385	385	383	379
Combustible fuels	363	381	385	385	383	379
Hydro	..	..	..	..	..	..
Nuclear	..	..	..	..	..	..
Other	..	..	..	..	..	..
Own use in electricity, CHP and heat plants	14	14	15	15	15	14
Net production	349	367	371	370	368	366
Imports	..	..	..	..	..	..
Exports	..	..	..	..	..	..
Losses	34	36	37	37	34	34
Consumption	315	331	333	333	334	332
Energy industries own use	..	..	..	..	..	..
By industry and construction	19	18	19	18	18	18
By transport	..	..	..	..	..	..
By households and other cons.	296	312	315	316	317	314
Net installed capacity	**Thousand kilowatts**					
Total main activity and autoproducer	76	76	76	86	86	88
Combustible fuels	76	76	76	86	86	88
Hydro	..	..	..	..	..	..
Nuclear	..	..	..	..	..	..
Other	..	..	..	..	..	..
Main activity	76	76	76	86	86	88
Combustible fuels	76	76	76	86	86	88
Hydro	..	..	..	..	..	..
Nuclear	..	..	..	..	..	..
Other	..	..	..	..	..	..
Combustible fuel input	**Terajoules**					
Gas-diesel oil	3076	3294	3350	3321	3306	3298
Total input	3076	3294	3350	3321	3306	3298
Total production	1307	1371	1387	1385	1379	1366
Estimated efficiency (% of production to input)	42	42	41	42	42	41

Statistics on electricity

St. Pierre-Miquelon

Item	2009	2010	2011	2012	2013	2014
Production, trade and consumption	Million kilowatt-hours					
Total main activity and autoproducer	45	45	45	43	47	48
Combustible fuels	44	44	44	43	46	47
Hydro	..	..	..	..	..	..
Nuclear	..	..	..	..	..	..
Other	1	1	1	1	*1	*1
Main activity	45	45	45	43	47	48
Combustible fuels	44	44	44	43	46	47
Hydro	..	..	..	..	..	..
Nuclear	..	..	..	..	..	..
Other	1	1	1	1	*1	*1
Own use in electricity, CHP and heat plants	0	0	0	0	0	0
Net production	45	45	45	43	47	48
Imports	..	..	..	..	..	..
Exports	..	..	..	..	..	..
Losses	3	3	2	3	*3	*3
Consumption	42	42	44	41	*43	*45
Energy industries own use	..	..	..	..	..	..
By industry and construction	..	..	..	..	..	..
By transport	..	..	..	..	..	..
By households and other cons.	42	42	44	41	*43	*45
Net installed capacity	Thousand kilowatts					
Total main activity and autoproducer	27	27	27	27	28	27
Combustible fuels	26	26	26	26	27	26
Hydro	..	..	..	..	..	..
Nuclear	..	..	..	..	..	..
Other	1	1	1	1	1	1
Main activity	27	27	27	27	28	27
Combustible fuels	26	26	26	26	27	26
Hydro	..	..	..	..	..	..
Nuclear	..	..	..	..	..	..
Other	1	1	1	1	1	1
Combustible fuel input	Terajoules					
Gas-diesel oil	*482	*482	*482	*464	*490	*516
Total input	*482	*482	*482	*464	*490	*516
Total production	159	159	159	153	164	171
Estimated efficiency (% of production to input)	33	33	33	33	34	33

193

Statistics on electricity

St. Vincent-Grenadines

Item	2009	2010	2011	2012	2013	2014
Production, trade and consumption	**Million kilowatt-hours**					
Total main activity and autoproducer	*140	*142	*145	*145	*139	*170
Combustible fuels	*116	*117	*114	*114	*108	*132
Hydro	*24	*24	*31	*31	*31	*37
Nuclear	..	..	..	..	..	..
Other	..	..	..	..	..	..
Main activity	*140	*142	*145	*145	*139	*170
Combustible fuels	*116	*117	*114	*114	*108	*132
Hydro	*24	*24	*31	*31	*31	*37
Nuclear	..	..	..	..	..	..
Other	..	..	..	..	..	..
Own use in electricity, CHP and heat plants	*5	*5	*4	*4	*4	*6
Net production	*135	*137	*141	*141	*135	*164
Imports	..	..	..	..	..	..
Exports	..	..	..	..	..	..
Losses	*11	*11	*11	9	*10	*12
Consumption	*124	*126	123	123	125	*158
Energy industries own use	..	..	..	..	..	..
By industry and construction	*6	*6	7	7	7	*9
By transport	..	..	..	..	..	..
By households and other cons.	*118	*120	116	116	119	*149
Net installed capacity	**Thousand kilowatts**					
Total main activity and autoproducer	*41	*41	*58	*58	*58	*58
Combustible fuels	*34	*34	*46	*46	*46	*46
Hydro	*7	*7	*13	*13	*13	*13
Nuclear	..	..	..	..	..	..
Other	..	..	..	..	..	..
Main activity	*41	*41	*58	*58	*58	*58
Combustible fuels	*34	*34	*46	*46	*46	*46
Hydro	*7	*7	*13	*13	*13	*13
Nuclear	..	..	..	..	..	..
Other	..	..	..	..	..	..
Combustible fuel input	**Terajoules**					
Gas-diesel oil	*1630	*1170	*991	*1421	*1125	*1124
Total input	*1630	*1170	*991	*1421	*1125	*1124
Total production	*418	*423	*411	*411	*390	*477
Estimated efficiency (% of production to input)	26	36	41	29	35	42

194

Statistics on electricity

State of Palestine

Item	2009	2010	2011	2012	2013	2014
Production, trade and consumption	Million kilowatt-hours					
Total main activity and autoproducer	501	473	569	461	536	337
Combustible fuels	501	473	569	461	536	337
Hydro	..	..	..	..	..	..
Nuclear	..	..	..	..	..	..
Other	..	..	..	..	..	..
Main activity	501	473	569	461	536	337
Combustible fuels	501	473	569	461	536	337
Hydro	..	..	..	..	..	..
Nuclear	..	..	..	..	..	..
Other	..	..	..	..	..	..
Own use in electricity, CHP and heat plants	0	0	0	0	0	0
Net production	501	473	569	461	536	337
Imports	3983	4159	4622	4909	4734	4938
Exports	..	..	..	..	..	..
Losses	897	695	1292	537	527	633
Consumption	3516	3280	3506	4846	4743	4642
Energy industries own use	..	..	..	..	..	..
By industry and construction	272	349	*299	*1026	405	527
By transport	41	..	..	..	..	..
By households and other cons.	3204	2931	3207	3820	4339	4115
Net installed capacity	Thousand kilowatts					
Total main activity and autoproducer	140	140	140	140	140	140
Combustible fuels	140	140	140	140	140	140
Hydro	..	..	..	..	..	..
Nuclear	..	..	..	..	..	..
Other	..	..	..	..	..	..
Main activity	140	140	140	140	140	140
Combustible fuels	140	140	140	140	140	140
Hydro	..	..	..	..	..	..
Nuclear	..	..	..	..	..	..
Other	..	..	..	..	..	..
Combustible fuel input	Terajoules					
Motor gasoline	310	758	89	257	354	337
Gas-diesel oil	4304	3298	5199	3904	4455	2989
Total input	4614	4056	5287	4161	4809	3325
Total production	1802	1704	2049	1660	1930	1212
Estimated efficiency (% of production to input)	39	42	39	40	40	36

Statistics on electricity

Sudan

Item	2009	2010	2011	2012	2013	2014
Production, trade and consumption				**Million kilowatt-hours**		
Total main activity and autoproducer	..	..	..	9510	10607	11376
Combustible fuels	..	..	..	2891	2290	2463
Hydro	..	..	..	6619	8317	8913
Nuclear	..	..	..	..	..	..
Other	..	..	..	..	..	..
Main activity	..	..	..	9436	10287	11376
Combustible fuels	..	..	..	2817	1970	2463
Hydro	..	..	..	6619	8317	8913
Nuclear	..	..	..	..	..	..
Other	..	..	..	..	..	..
Own use in electricity, CHP and heat plants	..	..	..	106	356	41
Net production	..	..	..	9404	10251	11335
Imports	..	..	..	..	..	..
Exports	..	..	..	..	..	..
Losses	..	..	..	1794	2356	1625
Consumption	..	..	..	7611	7895	9710
Energy industries own use	..	..	..	..	..	..
By industry and construction	..	..	..	1125	1279	1628
By transport	..	..	..	..	..	..
By households and other cons.	..	..	..	6486	6616	8082
Net installed capacity				**Thousand kilowatts**		
Total main activity and autoproducer	..	..	..	2877	3117	*3117
Combustible fuels	..	..	..	1287	1287	*1287
Hydro	..	..	..	1590	1830	*1830
Nuclear	..	..	..	..	..	..
Other	..	..	..	..	..	..
Main activity	..	..	..	2877	3117	*3117
Combustible fuels	..	..	..	1287	1287	*1287
Hydro	..	..	..	1590	1830	*1830
Nuclear	..	..	..	..	..	..
Other	..	..	..	..	..	..
Combustible fuel input				**Terajoules**		
Gas-diesel oil	..	..	..	10922	11051	11395
Fuel oil	..	..	..	14423	*9211	15069
Other oil products	..	..	..	*603	*2010	0
Total input	..	..	..	25948	*22272	26464
Total production	..	..	..	10408	8244	8867
Estimated efficiency (% of production to input)	..	..	..	40	37	34

Statistics on electricity

Suriname

Item	2009	2010	2011	2012	2013	2014
Production, trade and consumption	Million kilowatt-hours					
Total main activity and autoproducer	1740	1723	1951	1972	2184	2180
Combustible fuels	671	755	1005	974	1703	1385
Hydro	1069	968	946	998	481	795
Nuclear	..	..	..	..	..	..
Other	..	..	..	..	..	..
Main activity	270	332	347	526	670	589
Combustible fuels	270	332	347	526	670	589
Hydro	..	..	..	..	..	..
Nuclear	..	..	..	..	..	..
Other	..	..	..	..	..	..
Own use in electricity, CHP and heat plants	30	29	33	34	37	35
Net production	1711	1694	1918	1939	2147	2145
Imports	..	..	..	..	..	..
Exports	..	..	..	..	..	..
Losses	157	147	176	178	197	190
Consumption	1571	1547	1742	1761	1951	1955
Energy industries own use	..	..	..	..	..	..
By industry and construction	830	811	887	866	962	946
By transport	..	..	..	..	..	..
By households and other cons.	741	737	855	896	988	1009
Net installed capacity	Thousand kilowatts					
Total main activity and autoproducer	389	410	427	427	427	427
Combustible fuels	200	221	238	238	238	238
Hydro	189	189	189	189	189	189
Nuclear	..	..	..	..	..	..
Other	..	..	..	..	..	..
Main activity	60	80	115	115	115	115
Combustible fuels	60	80	115	115	115	115
Hydro	..	..	..	..	..	..
Nuclear	..	..	..	..	..	..
Other	..	..	..	..	..	..
Combustible fuel input	Terajoules					
Gas-diesel oil	490	675	710	856	1028	731
Fuel oil	12504	13849	11672	12290	12972	13756
Natural gas	185	183	172	175	179	149
Total input	13179	14707	12553	13321	14179	14636
Total production	2417	2718	3618	3508	6132	4984
Estimated efficiency (% of production to input)	18	18	29	26	43	34

197

Statistics on electricity

Swaziland

Item	2009	2010	2011	2012	2013	2014
Production, trade and consumption	**Million kilowatt-hours**					
Total main activity and autoproducer	*518	560	654	*622	*590	567
Combustible fuels	*272	*272	*321	*350	*350	265
Hydro	246	288	333	272	240	302
Nuclear	..	..	..	..	..	..
Other	..	..	..	..	..	..
Main activity	246	288	333	280	248	302
Combustible fuels	1	0	0	8	*8	0
Hydro	246	288	333	272	240	302
Nuclear	..	..	..	..	..	..
Other	..	..	..	..	..	..
Own use in electricity, CHP and heat plants	16	14	15	15	15	0
Net production	*502	546	639	*607	*575	567
Imports	923	909	806	813	822	860
Exports	..	..	..	..	..	..
Losses	188	168	161	175	169	171
Consumption	1238	1288	1283	1246	1227	1303
Energy industries own use	..	..	..	..	..	285
By industry and construction	406	390	360	338	319	336
By transport	..	..	..	..	..	..
By households and other cons.	832	898	923	908	909	682
Net installed capacity	**Thousand kilowatts**					
Total main activity and autoproducer	*152	*153	*159	*164	*164	*173
Combustible fuels	*89	*89	*95	*100	*100	*100
Hydro	64	64	64	64	64	73
Nuclear	..	..	..	..	..	..
Other	..	..	..	..	..	..
Main activity	69	70	70	70	70	79
Combustible fuels	10	10	10	10	10	10
Hydro	60	60	60	60	60	69
Nuclear	..	..	..	..	..	..
Other	..	..	..	..	..	..
Combustible fuel input	**Terajoules**					
Hard Coal	*449	*297	*253	*338	*338	*297
Gas-diesel oil	*455	*455	*473	*473	*473	0
Bagasse	*13722	*13514	*14232	*13606	*13606	*13637
Black liquor	0	0	0	0	0	0
Total input	*14626	*14266	*14958	*14417	*14417	*13934
Total production	*980	*978	*1156	*1258	*1260	954
Estimated efficiency (% of production to input)	7	7	8	9	9	7

Statistics on electricity

Sweden

Item	2009	2010	2011	2012	2013	2014
Production, trade and consumption	**Million kilowatt-hours**					
Total main activity and autoproducer	136729	148563	150376	166562	153166	153662
Combustible fuels	16087	20723	17256	16283	15336	13632
Hydro	65977	66501	66556	79058	61496	63872
Nuclear	52173	57828	60475	64037	66457	64877
Other	2492	3511	6089	7184	9877	11281
Main activity	131140	141957	144361	160212	147312	147865
Combustible fuels	10514	14131	11256	9950	9492	7847
Hydro	65961	66487	66541	79041	61486	63860
Nuclear	52173	57828	60475	64037	66457	64877
Other	2492	3511	6089	7184	9877	11281
Own use in electricity, CHP and heat plants	3386	3284	3440	3732	3627	3706
Net production	133343	145279	146936	162830	149539	149956
Imports	13765	14931	12481	11682	12674	13852
Exports	9080	12853	19714	31255	22676	29475
Losses	9905	10585	10569	10961	10003	7334
Consumption	126503	135168	127802	130503	128075	125195
Energy industries own use	3117	3951	3174	3217	3059	3004
By industry and construction	51431	54386	53802	54077	52014	50730
By transport	2438	2404	2640	2685	2750	2615
By households and other cons.	69517	74427	68186	70524	70252	68846
Net installed capacity	**Thousand kilowatts**					
Total main activity and autoproducer	35285	36454	35227	37843	37915	39930
Combustible fuels	8337	8715	6546	8362	7776	9270
Hydro	16652	16732	16577	16414	16494	15996
Nuclear	8839	8977	9323	9436	9408	9507
Other	1457	2030	2781	3631	4237	5157
Main activity	34046	35334	33825	36602	36712	38727
Combustible fuels	7101	7598	5147	7124	6582	8076
Hydro	16649	16729	16574	16411	16485	15987
Nuclear	8839	8977	9323	9436	9408	9507
Other	1457	2030	2781	3631	4237	5157
Combustible fuel input	**Terajoules**					
Hard Coal	9900	12784	10500	9167	12813	8423
Peat	12212	12486	11115	8884	7588	5261
Fuel oil	7838	13413	5656	5939	3596	3192
Natural gas	24910	32541	21808	14942	14963	7695
Blast furnace gas	4814	8232	7235	5630	5122	5749
Fuelwood	*115429	*140000	*103349	*131404	*128628	*121842
Municipal waste	35988	43800	45724	49735	53494	56454
Industrial waste	846	868	1583	1283	1580	1360
Other recovered gases	943	727	256	1228	1169	1151
Other liquid biofuels	3151	1836	27	27	1562	658
Others	2528	6074	3850	2242	1697	911
Total input	*218560	*272761	211103	*230481	*232211	*212695
Total production	57913	74603	62122	58619	55210	49075
Estimated efficiency (% of production to input)	26	27	29	25	24	23

Statistics on electricity

Switzerland

Item	2009	2010	2011	2012	2013	2014
Production, trade and consumption			Million kilowatt-hours			
Total main activity and autoproducer	68453	67815	64629	69893	70312	71767
Combustible fuels	3187	3531	3567	3739	3720	3566
Hydro	37507	37825	34133	40305	39968	39701
Nuclear	27686	26339	26710	25441	25990	27557
Other	73	120	219	408	634	943
Main activity	62867	62220	59215	63828	63915	64902
Combustible fuels	541	820	785	812	700	457
Hydro	34617	35024	31650	37487	37135	36787
Nuclear	27686	26339	26710	25441	25990	27557
Other	23	37	70	88	90	101
Own use in electricity, CHP and heat plants	2098	1686	1666	1683	1701	1759
Net production	66355	66129	62963	68210	68611	70008
Imports	31368	33401	34824	31549	29874	28530
Exports	33525	32881	32237	33749	32270	34021
Losses	4181	4370	4485	4627	4760	4696
Consumption	60006	62266	61052	61370	61445	59812
Energy industries own use	2523	2494	2466	2411	2132	2355
By industry and construction	18208	19268	19205	19028	18767	18019
By transport	3064	3164	3061	3094	3142	3069
By households and other cons.	36211	37340	36320	36837	37404	36369
Net installed capacity			Thousand kilowatts			
Total main activity and autoproducer	17775	18087	18234	18584	18932	19165
Combustible fuels	919	944	918	1018	1021	993
Hydro	13521	13723	13769	13802	13817	13743
Nuclear	3238	3253	3278	3278	3278	3308
Other	97	167	269	486	816	1121
Main activity	16500	16802	16844	16838	16861	16813
Combustible fuels	253	265	240	291	291	263
Hydro	12991	13242	13280	13220	13232	13182
Nuclear	3238	3253	3278	3278	3278	3308
Other	18	42	46	49	60	60
Combustible fuel input			Terajoules			
Biogas	1191	1311	1451	1660	1815	1898
Gas-diesel oil	473	344	301	387	301	301
Fuel oil	121	40	0	0	0	..
Liquefied petroleum gas	47	47	47	47	47	47
Natural gas	7651	10357	8791	7962	6190	3321
Fuelwood	*3742	*3576	*4579	*5614	*6161	*6041
Municipal waste	36734	38350	39768	40768	40544	41292
Industrial waste	2164	1997	1804	1884	1888	1968
Total input	52124	56023	56741	58322	56946	54868
Total production	11473	12712	12841	13460	13392	12838
Estimated efficiency (% of production to input)	22	23	23	23	24	23

Syrian Arab Republic

Item	2009	2010	2011	2012	2013	2014
Production, trade and consumption	**Million kilowatt-hours**					
Total main activity and autoproducer	43308	46413	42025	31188	25933	21726
Combustible fuels	41442	43821	38742	27951	22933	18726
Hydro	1866	2592	3283	3237	3000	3000
Nuclear	..	..	..	..	..	..
Other	..	..	..	..	..	..
Main activity	42219	45331	40965	30436	25435	21229
Combustible fuels	40353	42739	37682	27199	22435	18229
Hydro	1866	2592	3283	3237	3000	3000
Nuclear	..	..	..	..	..	..
Other	..	..	..	..	..	..
Own use in electricity, CHP and heat plants	4948	5303	4802	3564	2963	2482
Net production	38360	41110	37223	27624	22970	19244
Imports	542	690	902	0	0	0
Exports	614	1043	1192	0	529	138
Losses	11103	7103	6461	5010	3636	3353
Consumption	27185	33654	30472	22614	18805	15753
Energy industries own use	..	..	..	..	..	..
By industry and construction	7427	11318	10248	7605	6324	5298
By transport	..	..	..	..	..	..
By households and other cons.	19758	22336	20224	15009	12481	10455
Net installed capacity	**Thousand kilowatts**					
Total main activity and autoproducer	8322	8206	8206	*8206	8206	*9603
Combustible fuels	7422	7356	7356	*7356	7356	*8109
Hydro	900	850	850	*850	850	1494
Nuclear	..	..	..	..	..	..
Other	..	..	..	..	..	..
Main activity	7138	7210	7210	*7210	7210	*8508
Combustible fuels	6238	6360	6360	*6360	6360	*7014
Hydro	900	850	850	*850	850	1494
Nuclear	..	..	..	..	..	..
Other	..	..	..	..	..	..
Combustible fuel input	**Terajoules**					
Gas-diesel oil	19479	19350	18963	13459	8901	8901
Fuel oil	197596	161438	152712	108393	64115	43874
Natural gas	213852	275817	242805	198995	163515	151174
Total input	430927	456605	414480	320847	236531	203949
Total production	149191	157756	139471	100624	82559	67414
Estimated efficiency (% of production to input)	35	35	34	31	35	33

Statistics on electricity

T.F.Yug.Rep. Macedonia

Item	2009	2010	2011	2012	2013	2014
Production, trade and consumption	**Million kilowatt-hours**					
Total main activity and autoproducer	6828	7260	6759	6262	6094	5303
Combustible fuels	5558	4829	5325	5218	4501	4082
Hydro	1270	2431	1433	1041	1584	1207
Nuclear	..	..	..	..	..	..
Other	..	..	1	3	9	14
Main activity	6827	7258	6757	6260	6092	5302
Combustible fuels	5557	4827	5323	5216	4499	4081
Hydro	1270	2431	1433	1041	1584	1207
Nuclear	..	..	..	..	..	..
Other	..	..	1	3	9	14
Own use in electricity, CHP and heat plants	494	433	475	465	422	327
Net production	6334	6827	6284	5797	5672	4976
Imports	1438	1420	2749	2741	2491	3074
Exports	0	0	73	72	62	115
Losses	1187	1279	1388	1296	1152	1069
Consumption	6585	6968	7572	7170	6949	6867
Energy industries own use	193	184	194	165	148	138
By industry and construction	1562	2010	2506	2210	2260	2206
By transport	27	20	17	17	17	19
By households and other cons.	4803	4754	4855	4778	4524	4504
Net installed capacity	**Thousand kilowatts**					
Total main activity and autoproducer	1599	1644	1647	1709	1734	1764
Combustible fuels	1046	1089	1089	1110	1110	1080
Hydro	553	555	556	595	617	632
Nuclear	..	..	..	..	..	..
Other	..	0	2	4	7	52
Main activity	1596	1641	1644	1706	1731	1761
Combustible fuels	1043	1086	1086	1107	1107	1077
Hydro	553	555	556	595	617	632
Nuclear	..	..	..	..	..	..
Other	..	0	2	4	7	52
Combustible fuel input	**Terajoules**					
Brown Coal	54706	53065	58284	55961	45809	43856
Gas-diesel oil	2	0	0	0	0	0
Fuel oil	3301	1560	710	935	1246	1663
Liquefied petroleum gas	0	0	0	0	0	0
Refinery gas	84	2	0	0	0	0
Natural gas	322	733	1216	3092	3443	1714
Fuelwood	0	0	0	0	0	0
Total input	58415	55359	60210	59989	50498	47233
Total production	20009	17384	19170	18785	16204	14695
Estimated efficiency (% of production to input)	34	31	32	31	32	31

Statistics on electricity

Tajikistan

Item	2009	2010	2011	2012	2013	2014
Production, trade and consumption	**Million kilowatt-hours**					
Total main activity and autoproducer	16166	16436	16234	16998	17115	16472
Combustible fuels	217	35	38	74	44	472
Hydro	15949	16401	16196	16924	17071	16000
Nuclear	..	..	..	..	..	..
Other	..	..	..	..	..	..
Main activity	16166	16436	16234	16998	17115	16472
Combustible fuels	217	35	38	74	44	472
Hydro	15949	16401	16196	16924	17071	16000
Nuclear	..	..	..	..	..	..
Other	..	..	..	..	..	..
Own use in electricity, CHP and heat plants	90	92	90	91	90	24
Net production	16076	16344	16144	16907	17025	16448
Imports	4304	432	172	114	117	33
Exports	4247	286	197	775	1061	1326
Losses	2099	2330	2271	2445	2528	2804
Consumption	13602	14160	13752	13777	13553	12333
Energy industries own use	66	67	66	66	66	63
By industry and construction	6143	7435	6450	6344	5496	4057
By transport	23	30	161	37	38	41
By households and other cons.	7370	6628	7075	7330	7953	8172
Net installed capacity	**Thousand kilowatts**					
Total main activity and autoproducer	5078	5105	5079	6168	5190	5190
Combustible fuels	319	318	324	314	318	318
Hydro	4759	4787	4755	5854	4872	4872
Nuclear	..	..	..	..	..	..
Other	..	..	..	..	..	..
Main activity	5078	5105	5079	6168	5190	5190
Combustible fuels	319	318	324	314	318	318
Hydro	4759	4787	4755	5854	4872	4872
Nuclear	..	..	..	..	..	..
Other	..	..	..	..	..	..
Combustible fuel input	**Terajoules**					
Natural gas	1618	719	728	933	787	3611
Total input	1618	719	728	933	787	3611
Total production	781	126	137	266	158	1699
Estimated efficiency (% of production to input)	48	18	19	29	20	47

Statistics on electricity

Thailand

Item	2009	2010	2011	2012	2013	2014
Production, trade and consumption	**Million kilowatt-hours**					
Total main activity and autoproducer	148390	159518	151595	166471	175531	180862
Combustible fuels	141230	153953	143632	158013	168397	173631
Hydro	7148	5537	7935	8431	5748	5540
Nuclear	..	..	..	..	..	..
Other	12	28	*28	*27	1386	1691
Main activity	133458	144382	136659	151337	145999	145972
Combustible fuels	126298	138817	128696	142879	139729	139849
Hydro	7148	5537	7935	8431	5748	5540
Nuclear	..	..	..	..	..	..
Other	12	28	*28	*27	522	583
Own use in electricity, CHP and heat plants	5284	5749	3514	5996	11644	11726
Net production	143106	153769	148081	160475	163887	169136
Imports	2439	7287	10774	10527	12569	12260
Exports	1560	1615	1335	1912	1794	2066
Losses	8776	10121	10833	9502	10340	10545
Consumption	135209	149320	*146687	*159588	164322	168785
Energy industries own use	284	373	*325	*360	..	..
By industry and construction	56386	63257	*67942	*72336	68606	73782
By transport	62	74	*64	*71	164	166
By households and other cons.	78477	85616	*78356	*86821	95552	94837
Net installed capacity	**Thousand kilowatts**					
Total main activity and autoproducer	46578	47456	48750	49806	52538	53472
Combustible fuels	43077	43939	45233	46289	47250	48238
Hydro	3488	3488	*3488	*3488	3418	3444
Nuclear	..	..	..	..	..	..
Other	13	29	*29	*29	1870	1790
Main activity	30607	31485	32779	33835	36564	37498
Combustible fuels	27106	27968	29262	30318	31276	32264
Hydro	3488	3488	*3488	*3488	3418	3444
Nuclear	..	..	..	..	..	..
Other	13	29	*29	*29	1870	1790
Combustible fuel input	**Terajoules**					
Hard Coal	115267	116956	161823	205187	223334	224205
Brown Coal	165929	167562	208369	203427	204993	182557
Biogas	1526	2611	4141	1564	7395	7564
Gas-diesel oil	903	1419	*1462	*1591	2537	1548
Fuel oil	6626	10140	*9534	*9130	12322	14665
Natural gas	943830	1043016	998589	1076831	976868	1027756
Fuelwood	*111220	*125954	*82066	*69314	*98537	*160849
Bagasse	*153098	*148112	*186482	*221809	*238965	*278054
Municipal waste	169	127	127	1338	1758	1902
Total input	1498568	1615897	1652593	1790191	1766709	1899099
Total production	508428	554231	517075	568847	606229	625072
Estimated efficiency (% of production to input)	34	34	31	32	34	33

Statistics on electricity

Timor-Leste

Item	2009	2010	2011	2012	2013	2014
Production, trade and consumption	Million kilowatt-hours					
Total main activity and autoproducer	132	137	140	126	295	349
Combustible fuels	132	137	140	126	295	349
Hydro	..	..	..	..	..	..
Nuclear	..	..	..	..	..	..
Other	..	..	..	..	..	..
Main activity	132	137	140	126	295	349
Combustible fuels	132	137	140	126	295	349
Hydro	..	..	..	..	..	..
Nuclear	..	..	..	..	..	..
Other	..	..	..	..	..	..
Own use in electricity, CHP and heat plants	0	0	0	0	0	0
Net production	132	137	140	126	295	349
Imports	..	..	..	..	..	..
Exports	..	..	..	..	..	..
Losses	*13	*14	*14	*13	*30	*35
Consumption	119	123	126	113	*265	*314
Energy industries own use	..	..	..	..	..	..
By industry and construction	..	..	..	..	..	..
By transport	..	..	..	..	..	..
By households and other cons.	119	123	126	113	*265	*314
Net installed capacity	Thousand kilowatts					
Total main activity and autoproducer	44	50	*50	51	138	286
Combustible fuels	44	50	*50	51	138	286
Hydro	..	..	..	..	..	..
Nuclear	..	..	..	..	..	..
Other	..	..	..	..	..	..
Main activity	44	50	*50	51	138	286
Combustible fuels	44	50	*50	51	138	286
Hydro	..	..	..	..	..	..
Nuclear	..	..	..	..	..	..
Other	..	..	..	..	..	..
Combustible fuel input	Terajoules					
Gas-diesel oil	*1591	*1634	*1677	*1505	*3440	*3440
Total input	*1591	*1634	*1677	*1505	*3440	*3440
Total production	474	492	504	454	1062	1256
Estimated efficiency (% of production to input)	30	30	30	30	31	37

Statistics on electricity

Togo

Item	2009	2010	2011	2012	2013	2014
Production, trade and consumption	**Million kilowatt-hours**					
Total main activity and autoproducer	262	320	320	399	360	358
Combustible fuels	43	135	119	234	286	238
Hydro	219	186	202	165	74	120
Nuclear	..	..	..	..	..	..
Other	..	..	..	..	..	..
Main activity	252	310	309	388	353	347
Combustible fuels	33	125	108	223	279	227
Hydro	219	186	202	165	74	120
Nuclear	..	..	..	..	..	..
Other	..	..	..	..	..	..
Own use in electricity, CHP and heat plants	3	8	6	11	13	6
Net production	259	313	314	389	347	351
Imports	615	628	696	679	773	857
Exports	..	..	..	..	..	..
Losses	147	162	163	164	183	186
Consumption	728	779	847	904	937	1023
Energy industries own use	..	..	..	..	..	..
By industry and construction	212	226	253	277	271	309
By transport	..	..	..	..	..	..
By households and other cons.	516	553	594	627	666	714
Net installed capacity	**Thousand kilowatts**					
Total main activity and autoproducer	135	169	201	227	227	227
Combustible fuels	68	102	134	160	160	160
Hydro	67	67	67	67	67	67
Nuclear	..	..	..	..	..	..
Other	..	..	..	..	..	..
Main activity	127	161	193	219	219	219
Combustible fuels	60	94	126	152	152	152
Hydro	67	67	67	67	67	67
Nuclear	..	..	..	..	..	..
Other	..	..	..	..	..	..
Combustible fuel input	**Terajoules**					
Gas-diesel oil	331	568	215	275	237	241
Fuel oil	..	400	533	1026	1382	598
Natural gas	..	..	*311	654	844	1481
Vegetal waste	58	58	72	72	74	76
Kerosene-type jet fuel	132	397	44	88	88	88
Total input	521	1422	1175	2115	2624	2483
Total production	155	485	427	843	1031	856
Estimated efficiency (% of production to input)	30	34	36	40	39	34

Statistics on electricity

Tonga

Item	2009	2010	2011	2012	2013	2014
Production, trade and consumption	**Million kilowatt-hours**					
Total main activity and autoproducer	55	52	53	52	*53	*55
Combustible fuels	55	52	53	52	*51	*52
Hydro	..	..	..	..	..	..
Nuclear	..	..	..	..	..	..
Other	..	..	..	..	*2	*2
Main activity	55	52	53	52	*53	*55
Combustible fuels	55	52	53	52	*51	*52
Hydro	..	..	..	..	..	..
Nuclear	..	..	..	..	..	..
Other	..	..	..	..	*2	*2
Own use in electricity, CHP and heat plants	2	2	2	2	*2	*2
Net production	53	50	51	50	*51	*53
Imports	..	..	..	..	..	..
Exports	..	..	..	..	..	..
Losses	8	7	7	6	5	5
Consumption	*45	*43	*45	*45	*46	*48
Energy industries own use	..	..	..	..	..	..
By industry and construction	..	..	..	..	..	..
By transport	..	..	..	..	..	..
By households and other cons.	*45	*43	*45	*45	*46	*48
Net installed capacity	**Thousand kilowatts**					
Total main activity and autoproducer	*12	14	*14	*14	*15	*16
Combustible fuels	*12	14	*14	*14	*14	*14
Hydro	..	..	..	..	..	..
Nuclear	..	..	..	..	..	..
Other	..	..	..	..	*1	*2
Main activity	*12	14	*14	*14	*15	*16
Combustible fuels	*12	14	*14	*14	*14	*14
Hydro	..	..	..	..	..	..
Nuclear	..	..	..	..	..	..
Other	..	..	..	..	*1	*2
Combustible fuel input	**Terajoules**					
Gas-diesel oil	490	464	477	0	0	0
Fuel oil	..	..	..	*465	*464	*464
Total input	490	464	477	*465	*464	*464
Total production	197	187	191	189	*185	*187
Estimated efficiency (% of production to input)	40	40	40	41	40	40

Statistics on electricity

Trinidad and Tobago

Item	2009	2010	2011	2012	2013	2014
Production, trade and consumption	**Million kilowatt-hours**					
Total main activity and autoproducer	7843	8485	8772	9132	9505	9891
Combustible fuels	7843	8485	8772	9132	9505	9891
Hydro	..	..	..	..	..	..
Nuclear	..	..	..	..	..	..
Other	..	..	..	..	..	..
Main activity	7808	8415	8700	9057	9427	9813
Combustible fuels	7808	8415	8700	9057	9427	9813
Hydro	..	..	..	..	..	..
Nuclear	..	..	..	..	..	..
Other	..	..	..	..	..	..
Own use in electricity, CHP and heat plants	287	309	319	332	346	360
Net production	7556	8176	8453	8800	9159	9531
Imports	..	..	..	..	..	..
Exports	..	..	..	..	..	..
Losses	291	265	243	239	234	228
Consumption	7295	7911	8211	8560	8926	9306
Energy industries own use	..	..	..	..	..	..
By industry and construction	4334	4761	4964	5175	5396	5626
By transport	..	..	..	..	..	..
By households and other cons.	2961	3150	3247	3385	3530	3680
Net installed capacity	**Thousand kilowatts**					
Total main activity and autoproducer	*1807	1796	2295	2577	2598	2598
Combustible fuels	*1807	1796	2295	2577	2598	2598
Hydro	..	..	..	..	..	..
Nuclear	..	..	..	..	..	..
Other	..	..	..	..	..	..
Main activity	*1611	1600	2099	2381	2402	2402
Combustible fuels	*1611	1600	2099	2381	2402	2402
Hydro	..	..	..	..	..	..
Nuclear	..	..	..	..	..	..
Other	..	..	..	..	..	..
Combustible fuel input	**Terajoules**					
Gas-diesel oil	215	215	215	215	215	215
Other oil products	683	0	0	0	..	..
Natural gas	109464	117916	122343	122678	122343	121136
Total input	110362	118131	122558	122893	122558	121351
Total production	28235	30546	31579	32875	34218	35608
Estimated efficiency (% of production to input)	26	26	26	27	28	29

Statistics on electricity

Tunisia

Item	2009	2010	2011	2012	2013	2014
Production, trade and consumption	**Million kilowatt-hours**					
Total main activity and autoproducer	15428	16369	16497	18059	18382	19024
Combustible fuels	15252	15705	16142	17537	17740	18254
Hydro	79	50	54	110	60	56
Nuclear	..	..	..	..	..	..
Other	97	614	301	412	582	714
Main activity	14556	15346	15801	17425	17637	18257
Combustible fuels	14380	15157	15638	17119	17219	17694
Hydro	79	50	54	110	60	56
Nuclear	..	..	..	..	..	..
Other	97	139	109	196	358	507
Own use in electricity, CHP and heat plants	474	551	583	647	641	653
Net production	14954	15818	15914	17412	17741	18371
Imports	192	141	131	384	177	536
Exports	152	122	161	426	233	625
Losses	1882	1874	2255	2735	2707	2842
Consumption	13102	13952	13611	14632	14970	15418
Energy industries own use	318	401	372	246	241	228
By industry and construction	4717	5121	4780	5113	5305	5445
By transport	96	78	87	96	93	93
By households and other cons.	7971	8352	8372	9177	9331	9652
Net installed capacity	**Thousand kilowatts**					
Total main activity and autoproducer	3805	3808	4217	4279	4539	4998
Combustible fuels	3690	3693	4101	4041	4270	4695
Hydro	62	62	62	62	62	62
Nuclear	..	..	..	..	..	..
Other	53	53	54	176	207	241
Main activity	3473	3599	4024	4117	4334	4792
Combustible fuels	3358	3484	3909	3882	4072	4497
Hydro	62	62	62	62	62	62
Nuclear	..	..	..	..	..	..
Other	53	53	53	173	200	233
Combustible fuel input	**Terajoules**					
Gas-diesel oil	43	43	129	43	43	43
Fuel oil	4929	40	0	0	646	3394
Natural gas	137389	156380	153722	165591	167834	173915
Total input	142361	156463	153851	165634	168523	177352
Total production	54907	56538	58111	63133	63864	65714
Estimated efficiency (% of production to input)	39	36	38	38	38	37

Statistics on electricity

Turkey

Item	2009	2010	2011	2012	2013	2014
Production, trade and consumption	**Million kilowatt-hours**					
Total main activity and autoproducer	194812	211208	229393	239496	240154	251963
Combustible fuels	156923	155828	171638	174872	171622	200171
Hydro	35958	51796	52338	57865	59420	40645
Nuclear	..	..	..	..	..	..
Other	1931	3584	5417	6759	9112	11147
Main activity	181314	198761	217150	226242	227083	247522
Combustible fuels	145020	144607	160637	163278	159822	196264
Hydro	34365	50572	51098	56207	58344	40254
Nuclear	..	..	..	..	..	..
Other	1929	3582	5415	6757	8917	11004
Own use in electricity, CHP and heat plants	8193	8162	11835	11789	11177	12514
Net production	186619	203046	217558	227707	228977	239449
Imports	812	1144	4556	5827	7429	7953
Exports	1546	1918	3645	2954	1227	2696
Losses	28991	30222	32369	35657	37134	37331
Consumption	156894	172050	186100	194923	198045	207375
Energy industries own use	1966	2036	2107	2027	1877	1933
By industry and construction	68504	77295	85874	90275	91384	95844
By transport	556	591	657	799	826	916
By households and other cons.	85868	92128	97462	101822	103958	108682
Net installed capacity	**Thousand kilowatts**					
Total main activity and autoproducer	44761	49524	52911	57059	64007	72893
Combustible fuels	29339	32279	33931	35027	38612	44576
Hydro	14553	15831	17137	19609	22289	24187
Nuclear	..	..	..	..	..	..
Other	869	1414	1843	2423	3106	4130
Main activity	41705	46381	49893	53858	60586	69493
Combustible fuels	26828	29681	31458	32371	35776	41740
Hydro	14009	15287	16593	19065	21745	23643
Nuclear	..	..	..	..	..	..
Other	868	1413	1842	2422	3065	4110
Combustible fuel input	**Terajoules**					
Hard Coal	165652	185519	272016	320640	319753	376489
Brown Coal	555158	528342	563988	475756	479314	576317
Biogas	1968	2846	3037	5387	8511	9741
Gas-diesel oil	4644	43	43	4988	4773	4386
Fuel oil	46702	22866	10100	10504	12605	22826
Natural gas	761660	793194	826120	832166	816124	917159
Coke-oven gas	6200	9387	9015	10257	9375	9521
Blast furnace gas	16629	17499	17968	18213	20524	27697
Fuelwood	*501	*603	*271	*534	*588	*1559
Industrial waste	1056	1496	1499	1567	1215	1403
Others	312	579	..	..	..	..
Total input	1560483	1562374	1704057	1680012	1672782	1947098
Total production	564923	560981	617897	629539	617839	720616
Estimated efficiency (% of production to input)	36	36	36	37	37	37

Statistics on electricity

Turkmenistan

Item	2009	2010	2011	2012	2013	2014
Production, trade and consumption	Million kilowatt-hours					
Total main activity and autoproducer	15983	16663	17220	17750	18870	20400
Combustible fuels	15980	16660	17220	17750	18870	20400
Hydro	3	3	0	0	..	..
Nuclear	..	..	..	..	..	..
Other	..	..	..	..	..	..
Main activity	15983	16663	17220	17750	18870	20400
Combustible fuels	15980	16660	17220	17750	18870	20400
Hydro	3	3	0	0	..	..
Nuclear	..	..	..	..	..	..
Other	..	..	..	..	..	..
Own use in electricity, CHP and heat plants	1211	1262	1301	1341	1426	1542
Net production	14772	15401	15919	16409	17444	18858
Imports	..	..	..	..	..	..
Exports	2100	2410	2550	2720	2850	3210
Losses	2094	2132	2190	2235	2385	2547
Consumption	10578	10859	11179	11454	12209	13101
Energy industries own use	1625	1668	1717	1759	1875	2012
By industry and construction	3230	3316	3414	3498	3729	4001
By transport	232	238	245	251	268	288
By households and other cons.	5491	5637	5803	5946	6337	6800
Net installed capacity	Thousand kilowatts					
Total main activity and autoproducer	2852	*3001	*3201	*3401	*3801	*4001
Combustible fuels	2851	*3000	*3200	*3400	*3800	*4000
Hydro	1	*1	*1	*1	*1	*1
Nuclear	..	..	..	..	..	..
Other	..	..	..	..	..	..
Main activity	2852	*3001	*3201	*3401	*3801	*4001
Combustible fuels	2851	*3000	*3200	*3400	*3800	*4000
Hydro	1	*1	*1	*1	*1	*1
Nuclear	..	..	..	..	..	..
Other	..	..	..	..	..	..
Combustible fuel input	Terajoules					
Natural gas	284874	326240	347452	359315	364390	371847
Total input	284874	326240	347452	359315	364390	371847
Total production	57528	59976	61992	63900	67932	73440
Estimated efficiency (% of production to input)	20	18	18	18	19	20

Statistics on electricity

Turks and Caicos Islands

Item	2009	2010	2011	2012	2013	2014
Production, trade and consumption	**Million kilowatt-hours**					
Total main activity and autoproducer	*208	*214	*214	*227	*227	*230
Combustible fuels	*208	*214	*214	*227	*227	*230
Hydro	..	..	..	..	..	..
Nuclear	..	..	..	..	..	..
Other	..	..	..	..	..	..
Main activity	*208	*214	*214	*227	*227	*230
Combustible fuels	*208	*214	*214	*227	*227	*230
Hydro	..	..	..	..	..	..
Nuclear	..	..	..	..	..	..
Other	..	..	..	..	..	..
Own use in electricity, CHP and heat plants	*7	*7	*7	*8	*8	*8
Net production	*201	*206	*207	*219	*219	*222
Imports	..	..	..	..	..	..
Exports	..	..	..	..	..	..
Losses	*10	*10	*11	*11	*11	*11
Consumption	*191	*196	*196	*208	*208	*211
Energy industries own use	..	..	..	..	..	..
By industry and construction	..	..	..	..	..	..
By transport	..	..	..	..	..	..
By households and other cons.	*191	*196	*196	*208	*208	*211
Net installed capacity	**Thousand kilowatts**					
Total main activity and autoproducer	54	57	65	76	76	79
Combustible fuels	54	57	65	76	76	79
Hydro	..	..	..	..	..	..
Nuclear	..	..	..	..	..	..
Other	..	..	..	..	..	..
Main activity	54	57	65	76	76	79
Combustible fuels	54	57	65	76	76	79
Hydro	..	..	..	..	..	..
Nuclear	..	..	..	..	..	..
Other	..	..	..	..	..	..
Combustible fuel input	**Terajoules**					
Gas-diesel oil	*2245	*2309	*2309	*2451	*2451	*2537
Total input	*2245	*2309	*2309	*2451	*2451	*2537
Total production	*747	*769	*770	*817	*817	*828
Estimated efficiency (% of production to input)	33	33	33	33	33	33

Statistics on electricity

Tuvalu

Item	2009	2010	2011	2012	2013	2014
Production, trade and consumption	Million kilowatt-hours					
Total main activity and autoproducer	5	5	5	*5	*5	*5
Combustible fuels	5	5	5	*5	*5	*5
Hydro	..	..	..	..	..	..
Nuclear	..	..	..	..	..	..
Other	..	..	..	..	..	..
Main activity	5	5	5	*5	*5	*5
Combustible fuels	5	5	5	*5	*5	*5
Hydro	..	..	..	..	..	..
Nuclear	..	..	..	..	..	..
Other	..	..	..	..	..	..
Own use in electricity, CHP and heat plants	0	0	0	0	0	0
Net production	5	5	5	*5	*5	*5
Imports	..	..	..	..	..	..
Exports	..	..	..	..	..	..
Losses	..	..	..	..	..	..
Consumption	5	5	5	*5	*5	*5
Energy industries own use	..	..	..	..	..	..
By industry and construction	..	..	..	..	..	..
By transport	..	..	..	..	..	..
By households and other cons.	5	5	5	*5	*5	*5
Net installed capacity	Thousand kilowatts					
Total main activity and autoproducer	*3	*3	*3	*3	3	3
Combustible fuels	*3	*3	*3	*3	3	3
Hydro	..	..	..	..	..	..
Nuclear	..	..	..	..	..	..
Other	..	..	..	..	..	..
Main activity	*3	*3	*3	*3	3	3
Combustible fuels	*3	*3	*3	*3	3	3
Hydro	..	..	..	..	..	..
Nuclear	..	..	..	..	..	..
Other	..	..	..	..	..	..
Combustible fuel input	Terajoules					
Gas-diesel oil	59	66	68	*69	*69	*69
Total input	59	66	68	*69	*69	*69
Total production	17	19	18	*18	*18	*18
Estimated efficiency (% of production to input)	28	29	27	27	27	27

Statistics on electricity

Uganda

Item	2009	2010	2011	2012	2013	2014
Production, trade and consumption	**Million kilowatt-hours**					
Total main activity and autoproducer	2186	*2456	*2588	*2640	*2861	3258
Combustible fuels	905	*851	897	*915	*915	697
Hydro	1281	*1605	*1691	*1725	*1946	2561
Nuclear	..	..	..	..	..	..
Other	..	..	..	..	..	..
Main activity	2127	*2375	*2503	*2553	*2774	3162
Combustible fuels	875	*823	867	*884	*884	657
Hydro	1252	*1553	*1636	*1669	*1890	2505
Nuclear	..	..	..	..	..	..
Other	..	..	..	..	..	..
Own use in electricity, CHP and heat plants	92	*97	*102	*113	*111	162
Net production	2093	*2359	*2486	*2526	*2750	3096
Imports	25	*29	*36	*37	47	33
Exports	81	*75	89	*91	105	125
Losses	*815	*791	*767	*782	*661	619
Consumption	*1240	*1521	*1681	*1724	2070	2417
Energy industries own use	..	..	..	..	..	..
By industry and construction	*755	*959	*1088	*1131	1349	1602
By transport	..	..	..	..	..	..
By households and other cons.	*485	*562	*593	*593	721	815
Net installed capacity	**Thousand kilowatts**					
Total main activity and autoproducer	504	548	628	728	838	883
Combustible fuels	166	186	186	136	136	178
Hydro	338	362	442	592	702	705
Nuclear	..	..	..	..	..	..
Other	..	..	..	..	..	..
Main activity	478	522	602	702	792	814
Combustible fuels	150	170	170	120	100	119
Hydro	328	352	432	582	692	695
Nuclear	..	..	..	..	..	..
Other	..	..	..	..	..	..
Combustible fuel input	**Terajoules**					
Gas-diesel oil	*7955	*8002	*8045	*8045	*8045	*8045
Fuel oil	..	117	117	117	*117	*117
Bagasse	*193	*193	*185	*208	*208	*347
Total input	*8148	*8313	*8348	*8371	*8371	*8510
Total production	3258	*3065	3229	*3294	*3294	2510
Estimated efficiency (% of production to input)	40	37	39	39	39	29

Statistics on electricity

Ukraine

Item	2009	2010	2011	2012	2013	2014
Production, trade and consumption	Million kilowatt-hours					
Total main activity and autoproducer	173645	188857	194947	198878	194377	182815
Combustible fuels	78742	86495	93634	97126	95487	83549
Hydro	11936	13158	10946	10994	14472	9318
Nuclear	82924	89152	90248	90137	83209	88389
Other	43	52	119	621	1209	1559
Main activity	170329	185300	191294	195105	191177	179888
Combustible fuels	75452	82966	90002	93373	92310	80644
Hydro	11910	13130	10925	10974	14450	9299
Nuclear	82924	89152	90248	90137	83209	88389
Other	43	52	119	621	1208	1556
Own use in electricity, CHP and heat plants	14474	15029	15312	15481	15293	14385
Net production	159171	173828	179635	183397	179084	168430
Imports	25	23	35	93	39	89
Exports	4294	4078	6327	11560	9929	8523
Losses	20692	21695	21261	21421	20714	19615
Consumption	134210	148078	152084	149817	148480	134981
Energy industries own use	14901	14053	12284	12149	10949	6594
By industry and construction	55860	65911	62890	63101	58584	54401
By transport	9144	8972	9887	9279	8690	8072
By households and other cons.	54305	59142	67023	65288	70257	65914
Net installed capacity	Thousand kilowatts					
Total main activity and autoproducer	54385	54567	54624	55001	55914	55843
Combustible fuels	35043	35186	35026	35129	35616	35335
Hydro	5421	5458	5469	5470	5489	5851
Nuclear	13835	13835	13835	13835	13835	13835
Other	86	88	294	567	974	822
Main activity	51126	51306	51486	51645	52578	52910
Combustible fuels	31790	31930	31894	31778	32290	32410
Hydro	5415	5453	5463	5465	5484	5847
Nuclear	13835	13835	13835	13835	13835	13835
Other	86	88	294	567	969	818
Combustible fuel input	Terajoules					
Hard Coal	747573	811579	901465	1001244	972367	855610
Gas-diesel oil	0	0	258	301	301	258
Fuel oil	41087	5010	3555	3717	3192	2020
Refinery gas	4109	4505	2772	743	594	446
Other oil products	2613	2653	2533	603	1769	1648
Natural gas	197370	290673	319166	288423	240341	204757
Coke-oven gas	23414	24466	26842	27085	26254	10037
Blast furnace gas	25735	26039	25004	24777	26079	23785
Fuelwood	0	0	..	..	..	6732
Other recovered gases	0	0	0	0	0	18592
Others	117	29	20	0	0	0
Total input	1042018	1164954	1281614	1346893	1270896	1123885
Total production	283471	311382	337082	349654	343753	300776
Estimated efficiency (% of production to input)	27	27	26	26	27	27

Statistics on electricity

United Arab Emirates

Item	2009	2010	2011	2012	2013	2014
Production, trade and consumption	**Million kilowatt-hours**					
Total main activity and autoproducer	85698	93949	99137	106222	109979	116528
Combustible fuels	85698	93949	99137	106222	109979	116528
Hydro	..	..	..	..	..	..
Nuclear	..	..	..	..	..	..
Other	..	..	..	..	..	..
Main activity	85698	93949	99137	106222	109979	116528
Combustible fuels	85698	93949	99137	106222	109979	116528
Hydro	..	..	..	..	..	..
Nuclear	..	..	..	..	..	..
Other	..	..	..	..	..	..
Own use in electricity, CHP and heat plants	1294	4362	3629	4768	4616	4843
Net production	84404	89587	95508	101454	105363	111685
Imports	0	0	49	16	0	0
Exports	0	0	36	34	23	0
Losses	6500	7013	7114	7363	7623	4843
Consumption	78384	84422	85549	88247	97717	106842
Energy industries own use	..	..	..	..	..	..
By industry and construction	8805	7591	9228	8140	13564	14355
By transport	..	..	..	..	..	..
By households and other cons.	69579	76831	76321	80107	84153	92487
Net installed capacity	**Thousand kilowatts**					
Total main activity and autoproducer	20565	23199	26086	27180	27374	28829
Combustible fuels	20565	23199	26076	27170	27314	28769
Hydro	..	..	..	..	..	..
Nuclear	..	..	..	..	..	..
Other	..	0	10	10	60	60
Main activity	20565	23199	26086	27180	27374	28829
Combustible fuels	20565	23199	26076	27170	27314	28769
Hydro	..	..	..	..	..	..
Nuclear	..	..	..	..	..	..
Other	..	0	10	10	60	60
Combustible fuel input	**Terajoules**					
Gas-diesel oil	*23478	*24897	*26703	*24725	*28982	21887
Fuel oil	*1899	*1939	*1939	*1980	*2141	1899
Natural gas	*1142240	*1167731	*1345717	*1240439	*1328899	*1209651
Total input	*1167617	*1194567	*1374359	*1267143	*1360022	*1233437
Total production	308513	338216	356893	382399	395924	419501
Estimated efficiency (% of production to input)	26	28	26	30	29	34

Statistics on electricity

United Kingdom

Item	2009	2010	2011	2012	2013	2014
Production, trade and consumption	Million kilowatt-hours					
Total main activity and autoproducer	376720	381771	367423	363582	359168	338925
Combustible fuels	289403	302618	273961	263734	250539	230342
Hydro	8915	6715	8585	8252	7606	8768
Nuclear	69098	62140	68980	70405	70607	63748
Other	9304	10298	15897	21191	30416	36067
Main activity	334395	340133	326858	322410	320465	296174
Combustible fuels	250778	263932	237539	227732	219387	198145
Hydro	7979	5853	7499	7136	6513	7518
Nuclear	69098	62140	68980	70405	70607	63748
Other	6540	8208	12840	17137	23958	26763
Own use in electricity, CHP and heat plants	16571	16110	16430	17971	17891	16519
Net production	360149	365661	350993	345611	341277	322406
Imports	6609	7144	8689	13743	17532	23244
Exports	3748	4481	2467	1872	3102	2723
Losses	28148	26612	27497	28334	26688	28011
Consumption	334861	341711	329715	329148	329018	314916
Energy industries own use	13114	12754	11773	11066	11891	11353
By industry and construction	99737	104654	102476	98299	97819	93526
By transport	4040	4251	4253	4262	4268	4260
By households and other cons.	217970	220052	211213	215521	215040	205777
Net installed capacity	Thousand kilowatts					
Total main activity and autoproducer	87457	93749	93671	95968	95105	97009
Combustible fuels	67772	72998	71121	70925	66679	64238
Hydro	4379	4387	4423	4439	4452	4467
Nuclear	10858	10865	10663	9946	9906	9937
Other	4448	5499	7464	10658	14068	18367
Main activity	79897	85975	84838	86301	83818	83056
Combustible fuels	61607	66565	64744	64515	60513	58362
Hydro	4203	4203	4221	4221	4221	4221
Nuclear	10858	10865	10663	9946	9906	9937
Other	3229	4342	5210	7619	9178	10536
Combustible fuel input	Terajoules					
Hard Coal	1012078	1065378	1068933	1396572	1280625	984384
Biogas	68728	72063	75431	78258	81506	84438
Fuel oil	30704	21897	13292	13615	7555	5939
Refinery gas	12177	16088	14306	9504	10989	10593
Natural gas	1293490	1357635	1112673	779554	742758	786223
Coke-oven gas	9453	9237	9363	8784	8361	7643
Blast furnace gas	22866	18955	19019	24742	30969	30715
Vegetal waste	*40700	*47054	*54744	*59959	*81356	*112443
Municipal waste	29225	29526	29024	34219	31824	32730
Industrial waste	6397	2439	13313	13313	12657	8422
Others	18680	9706	4011	5930	5442	4644
Total input	2544498	2649977	2414108	2424449	2294041	2068174
Total production	1041851	1089425	986260	949442	901940	829231
Estimated efficiency (% of production to input)	41	41	41	39	39	40

217

Statistics on electricity

United Rep. of Tanzania

Item	2009	2010	2011	2012	2013	2014
Production, trade and consumption	**Million kilowatt-hours**					
Total main activity and autoproducer	4741	5274	5093	5589	5941	6219
Combustible fuels	2096	2566	3091	3809	4209	3611
Hydro	2640	2701	1993	1767	1717	2590
Nuclear	..	..	..	..	..	..
Other	5	7	9	13	15	18
Main activity	4736	5267	5084	5576	5926	6201
Combustible fuels	2096	2566	3091	3809	4209	3611
Hydro	2640	2701	1993	1767	1717	2590
Nuclear	..	..	..	..	..	..
Other	..	..	..	..	..	..
Own use in electricity, CHP and heat plants	0	0	0	0	0	0
Net production	4741	5274	5093	5589	5941	6219
Imports	65	57	55	61	59	59
Exports	..	..	..	..	..	..
Losses	1686	1046	1159	1048	1140	1098
Consumption	3121	4099	4004	4454	4851	4994
Energy industries own use	5	7	9	13	15	18
By industry and construction	703	1036	1005	1089	1223	1270
By transport	..	..	..	..	..	..
By households and other cons.	2413	3056	2990	3352	3613	3706
Net installed capacity	**Thousand kilowatts**					
Total main activity and autoproducer	1011	1112	1113	1114	1115	1115
Combustible fuels	*446	*546	*546	*546	*546	*546
Hydro	562	562	562	562	562	562
Nuclear	..	..	..	..	..	..
Other	*3	*4	*5	*6	*7	*7
Main activity	988	1088	1088	1088	1088	1088
Combustible fuels	*426	*526	*526	*526	*526	*526
Hydro	562	562	562	562	562	562
Nuclear	..	..	..	..	..	..
Other	..	..	..	..	..	..
Combustible fuel input	**Terajoules**					
Gas-diesel oil	860	946	2451	3182	8428	5160
Fuel oil	566	1697	4040	8444	8201	7676
Natural gas	21708	25163	28187	31751	31411	28623
Bagasse	..	324	450	342	378	396
Total input	23134	28130	35128	43719	48418	41855
Total production	7546	9238	11128	13712	15152	13000
Estimated efficiency (% of production to input)	33	33	32	31	31	31

Statistics on electricity

United States

Item	2009	2010	2011	2012	2013	2014
Production, trade and consumption	**Million kilowatt-hours**					
Total main activity and autoproducer	4188214	4378422	4349571	4290660	4306371	4339210
Combustible fuels	2965173	3135578	3037683	3016908	2985891	2995556
Hydro	298410	286333	344679	298287	290113	281527
Nuclear	830210	838931	821405	801129	822004	830584
Other	94421	117580	145804	174336	208363	231543
Main activity	4039013	4214470	4185002	4116563	4128386	4166530
Combustible fuels	2820121	2976980	2880241	2854720	2822349	2835071
Hydro	296433	284550	342838	295887	286579	280195
Nuclear	830210	838931	821405	801129	822004	830584
Other	92249	114009	140518	164827	197454	220680
Own use in electricity, CHP and heat plants	209455	221165	215857	213509	212970	216819
Net production	3978759	4157257	4133714	4077151	4093401	4122391
Imports	52191	45083	52301	59257	70355	66511
Exports	18138	19107	15038	11995	11353	13298
Losses	260708	260999	259528	268753	255322	255322
Consumption	3752104	3922234	3911449	3855660	3897081	3920282
Energy industries own use	107403	135615	134448	128886	129517	132489
By industry and construction	797320	826432	851555	845911	846485	821040
By transport	6323	6417	6559	6841	7234	7605
By households and other cons.	2841058	2953770	2918887	2874022	2913845	2959148
Net installed capacity	**Thousand kilowatts**					
Total main activity and autoproducer	1026869	1041007	1055279	1067878	1065293	1073438
Combustible fuels	786050	793551	798833	793352	787297	787407
Hydro	100678	101023	100944	101106	101589	102162
Nuclear	101004	101167	101419	101885	99240	98569
Other	39137	45266	54083	71535	77167	85300
Main activity	996200	1008650	1020464	1031417	1027776	1036351
Combustible fuels	757555	764431	769510	764085	758052	758205
Hydro	100320	100661	100014	100442	100911	101856
Nuclear	101004	101167	101419	101885	99240	98569
Other	37321	42391	49521	65005	69573	77721
Combustible fuel input	**Terajoules**					
Hard Coal	9476906	10302749	9238760	7847311	8070925	8409391
Brown Coal	9175759	9374200	9197146	8305577	8624213	8268422
Biogas	107375	114408	122381	135501	146036	162387
Gas-diesel oil	74046	81614	65059	53922	56846	84452
Fuel oil	179578	151338	92193	76841	77204	95263
Petroleum coke	145893	150150	142643	105658	135785	124475
Natural gas	8033275	8610217	8828413	10578971	9626697	9602281
Animal waste	..	..	*141347	*104847	*149233	*171907
Vegetal waste	*235157	*257378	*250055	*330462	*289867	*312969
Municipal waste	274202	270201	277175	274975	274962	274254
Others	171696	164018	167178	181804	174698	180498
Total input	27873887	29476273	28522350	27995868	27626466	27686300
Total production	10674623	11288081	10935659	10860869	10749208	10784002
Estimated efficiency (% of production to input)	38	38	38	39	39	39

Statistics on electricity

United States Virgin Is.

Item	2009	2010	2011	2012	2013	2014
Production, trade and consumption	**Million kilowatt-hours**					
Total main activity and autoproducer	*921	958	951	873	821	*772
Combustible fuels	*921	958	951	872	820	*766
Hydro	..	..	..	..	..	..
Nuclear	..	..	..	..	..	..
Other	..	..	*1	*1	*1	*6
Main activity	*873	909	911	872	820	*766
Combustible fuels	*873	909	911	872	820	*766
Hydro	..	..	..	..	..	..
Nuclear	..	..	..	..	..	..
Other	..	..	..	..	..	..
Own use in electricity, CHP and heat plants	*70	86	85	86	80	*71
Net production	*850	872	866	787	741	*701
Imports	..	..	..	..	..	..
Exports	..	..	..	..	..	..
Losses	*68	*69	*70	*63	*60	*60
Consumption	772	803	796	724	681	641
Energy industries own use	..	..	..	..	..	..
By industry and construction	349	358	338	319	309	291
By transport	..	..	..	..	..	..
By households and other cons.	424	445	458	405	371	350
Net installed capacity	**Thousand kilowatts**					
Total main activity and autoproducer	*323	*323	345	331	317	321
Combustible fuels	*323	*323	345	330	316	316
Hydro	..	..	..	..	..	..
Nuclear	..	..	..	..	..	..
Other	..	..	*1	*1	*1	*5
Main activity	*308	*308	330	330	316	316
Combustible fuels	*308	*308	330	330	316	316
Hydro	..	..	..	..	..	..
Nuclear	..	..	..	..	..	..
Other	..	..	..	..	..	..
Combustible fuel input	**Terajoules**					
Total input	..	..	..	..	..	..
Total production	*3315	3447	3422	3140	2952	*2759
Estimated efficiency (% of production to input)	..	..	..	..	..	..

Statistics on electricity

Uruguay

Item	2009	2010	2011	2012	2013	2014
Production, trade and consumption	Million kilowatt-hours					
Total main activity and autoproducer	8666	10732	10345	10601	11657	13008
Combustible fuels	3564	2255	3755	5062	3309	2623
Hydro	5060	8407	6479	5421	8206	9649
Nuclear	..	..	..	..	..	..
Other	42	70	111	118	143	736
Main activity	7903	9903	9535	9734	10730	11728
Combustible fuels	2801	1426	2945	4197	2384	1349
Hydro	5060	8407	6479	5421	8206	9649
Nuclear	..	..	..	..	..	..
Other	42	70	111	117	140	730
Own use in electricity, CHP and heat plants	185	212	209	282	325	361
Net production	8481	10521	10135	10319	11332	12647
Imports	1468	387	477	742	0	0
Exports	243	711	19	20	206	1267
Losses	1136	1212	1288	1292	1281	1252
Consumption	8638	9060	9368	9646	9956	10239
Energy industries own use	68	75	60	66	106	107
By industry and construction	2405	2530	2526	2477	2625	2906
By transport	..	..	..	..	..	..
By households and other cons.	6165	6455	6782	7103	7225	7226
Net installed capacity	Thousand kilowatts					
Total main activity and autoproducer	2620	2690	2701	2911	3288	3713
Combustible fuels	1051	1111	1119	1320	1689	1690
Hydro	1538	1538	1538	1538	1538	1538
Nuclear	..	..	..	..	..	..
Other	31	41	44	53	61	485
Main activity	2427	2437	2437	2644	2844	3264
Combustible fuels	869	869	869	1069	1268	1268
Hydro	1538	1538	1538	1538	1538	1538
Nuclear	..	..	..	..	..	..
Other	20	30	30	37	38	458
Combustible fuel input	Terajoules					
Gas-diesel oil	20176	5057	15029	20124	9985	2928
Fuel oil	9942	6064	11502	18834	9235	4444
Natural gas	173	800	815	71	9	20
Fuelwood	29	416	92	338	479	95
Bagasse	32	36	76	54	82	100
Vegetal waste	0	223	583	2100	2165	2605
Black liquor	4222	4525	4554	4612	4956	6774
Total input	34575	17120	32651	46134	26910	16967
Total production	12830	8118	13517	18224	11911	9443
Estimated efficiency (% of production to input)	37	47	41	40	44	56

Statistics on electricity

Uzbekistan

Item	2009	2010	2011	2012	2013	2014
Production, trade and consumption	**Million kilowatt-hours**					
Total main activity and autoproducer	49950	51700	52400	52500	54200	55400
Combustible fuels	40620	40854	42160	41290	42640	43570
Hydro	9330	10846	10240	11210	11560	11830
Nuclear	..	..	..	..	..	..
Other	..	..	..	..	..	..
Main activity	49760	51509	52203	52307	54001	55196
Combustible fuels	40430	40663	41963	41097	42441	43366
Hydro	9330	10846	10240	11210	11560	11830
Nuclear	..	..	..	..	..	..
Other	..	..	..	..	..	..
Own use in electricity, CHP and heat plants	2845	2945	2985	2992	3089	3157
Net production	47105	48755	49415	49508	51111	52243
Imports	11916	12333	12500	12524	12930	13216
Exports	11831	12245	12411	12435	12838	13122
Losses	4410	4563	4623	4628	4778	4884
Consumption	42780	44280	44881	44969	46425	47453
Energy industries own use	1435	1485	1505	1508	1557	1591
By industry and construction	15835	16390	16612	16644	17183	17563
By transport	1360	1407	1426	1429	1476	1509
By households and other cons.	24150	24998	25338	25388	26209	26790
Net installed capacity	**Thousand kilowatts**					
Total main activity and autoproducer	*11580	*11580	*12400	*12722	*12722	*12722
Combustible fuels	*9870	*9870	*10569	*10867	*10867	*10867
Hydro	*1710	*1710	*1831	*1855	*1855	*1855
Nuclear	..	..	..	..	..	..
Other	..	..	..	..	..	..
Main activity	*11425	*11425	*12245	*12565	*12565	*12565
Combustible fuels	*9715	*9715	*10414	*10710	*10710	*10710
Hydro	*1710	*1710	*1831	*1855	*1855	*1855
Nuclear	..	..	..	..	..	..
Other	..	..	..	..	..	..
Combustible fuel input	**Terajoules**					
Brown Coal	37133	38437	38965	39038	40299	41178
Gas-diesel oil	43	43	43	0	0	..
Fuel oil	12484	9090	6504	4606	3353	2424
Natural gas	532848	539232	560347	550188	569848	583441
Total input	582508	586802	605859	593832	613500	627043
Total production	146232	147074	151776	148644	153504	156852
Estimated efficiency (% of production to input)	25	25	25	25	25	25

Statistics on electricity

Vanuatu

Item	2009	2010	2011	2012	2013	2014
Production, trade and consumption	*Million kilowatt-hours*					
Total main activity and autoproducer	69	69	*68	67	*68	*67
Combustible fuels	56	56	*57	57	*58	*56
Hydro	7	8	6	7	*7	*7
Nuclear	..	..	..	..	..	..
Other	6	5	*5	3	*3	*3
Main activity	69	69	*68	67	*68	*67
Combustible fuels	56	56	*57	57	*58	*56
Hydro	7	8	6	7	*7	*7
Nuclear	..	..	..	..	..	..
Other	6	5	*5	3	*3	*3
Own use in electricity, CHP and heat plants	2	2	2	2	*2	*2
Net production	67	67	*66	65	*66	*65
Imports	..	..	..	..	..	..
Exports	..	..	..	..	..	..
Losses	3	3	3	3	3	3
Consumption	57	61	63	61	62	61
Energy industries own use	..	..	..	..	..	..
By industry and construction	..	..	..	..	..	..
By transport	..	..	..	..	..	..
By households and other cons.	57	61	63	61	62	61
Net installed capacity	*Thousand kilowatts*					
Total main activity and autoproducer	28	29	31	31	31	31
Combustible fuels	25	25	27	27	27	27
Hydro	1	1	1	1	1	1
Nuclear	..	..	..	..	..	..
Other	2	3	3	3	3	3
Main activity	28	29	31	31	31	31
Combustible fuels	25	25	27	27	27	27
Hydro	1	1	1	1	1	1
Nuclear	..	..	..	..	..	..
Other	2	3	3	3	3	3
Combustible fuel input	*Terajoules*					
Gas-diesel oil	550	535	*538	*538	*542	*533
Other liquid biofuels	4	6	*27	*55	*82	*82
Total input	554	541	*565	*592	*624	*615
Total production	200	201	*205	205	*207	*203
Estimated efficiency (% of production to input)	36	37	36	35	33	33

223

Statistics on electricity

Venezuela (Bolivar. Rep.)

Item	2009	2010	2011	2012	2013	2014
Production, trade and consumption	**Million kilowatt-hours**					
Total main activity and autoproducer	119580	118370	122059	121653	123160	127732
Combustible fuels	33618	41590	38389	39645	39615	40538
Hydro	85962	76780	83670	82008	83545	87194
Nuclear	..	..	..	..	..	..
Other	..	..	..	..	..	..
Main activity	117905	116655	120533	120194	121623	126256
Combustible fuels	31943	39875	36863	38186	38078	39062
Hydro	85962	76780	83670	82008	83545	87194
Nuclear	..	..	..	..	..	..
Other	..	..	..	..	..	..
Own use in electricity, CHP and heat plants	975	4950	5275	5519	5677	5854
Net production	118605	113420	116784	116134	117483	121878
Imports	260	0	249	480	715	..
Exports	633	..	..	..	..	..
Losses	32564	22880	24133	25113	25627	46041
Consumption	83081	83770	87015	89987	99038	79576
Energy industries own use	2213	2084	2144	2114	1830	1847
By industry and construction	35662	36858	38094	39371	49445	33461
By transport	290	275	290	302	308	248
By households and other cons.	44916	44553	46487	48200	47455	44020
Net installed capacity	**Thousand kilowatts**					
Total main activity and autoproducer	24846	26157	27008	28832	31595	31765
Combustible fuels	10224	11535	12386	14177	16662	16748
Hydro	14622	14622	14622	14622	14880	14963
Nuclear	..	..	..	..	..	..
Other	..	..	..	33	53	53
Main activity	23708	24854	25705	27529	30292	30462
Combustible fuels	9086	10232	11083	12874	15359	15445
Hydro	14622	14622	14622	14622	14880	14963
Nuclear	..	..	..	..	..	..
Other	..	..	..	33	53	53
Combustible fuel input	**Terajoules**					
Gas-diesel oil	115025	143706	150199	199262	187652	194618
Fuel oil	59873	62458	52399	37936	36118	25654
Natural gas	218829	284506	253200	300446	254992	273413
Total input	393727	490670	455798	537644	478762	493685
Total production	121025	149724	138200	142722	142614	145937
Estimated efficiency (% of production to input)	31	31	30	27	30	30

Statistics on electricity

Viet Nam

Item	2009	2010	2011	2012	2013	2014
Production, trade and consumption	Million kilowatt-hours					
Total main activity and autoproducer	83185	94885	105219	118940	128549	145730
Combustible fuels	53194	66780	63944	65676	70287	83950
Hydro	29981	28055	41188	53172	58170	61480
Nuclear	..	..			..	..
Other	10	50	87	*92	*92	*300
Main activity	80720	91772	101586	115237	124546	141548
Combustible fuels	50729	63667	60311	61976	66287	79850
Hydro	29981	28055	41188	53171	58169	61400
Nuclear	..	..	..		..	..
Other	10	50	87	*90	*90	*298
Own use in electricity, CHP and heat plants	2017	3065	3355	*4003	*4549	*5767
Net production	81168	91820	101864	114937	124000	139963
Imports	199	299	301	159	221	2053
Exports	11	57	79	29	178	880
Losses	7985	*9650	10060	*11500	*11500	*12509
Consumption	76913	85669	93523	104255	115283	128627
Energy industries own use	..	..	..	..	..	..
By industry and construction	39919	44524	50328	55568	61277	69524
By transport	..	..	..	..	..	..
By households and other cons.	36994	41145	43195	48687	54006	59103
Net installed capacity	Thousand kilowatts					
Total main activity and autoproducer	*17458	*22881	*25731	*28240	*28240	*34080
Combustible fuels	*11950	*16950	*17700	*17700	*17700	*18368
Hydro	5500	5900	*8000	*10503	*10503	*15603
Nuclear	..	..	..	..	..	..
Other	8	31	*31	*37	*37	*109
Main activity	*15508	*20931	*23031	*25531	*25531	*30398
Combustible fuels	*10000	*15000	*15000	*15000	*15000	*15000
Hydro	5500	5900	*8000	*10500	*10500	*15295
Nuclear	..	..	..	..	..	..
Other	8	31	*31	*31	*31	*103
Combustible fuel input	Terajoules					
Hard Coal	154064	210076	254835	266675	321187	400645
Gas-diesel oil	946	10922	7310	1333	1548	1806
Fuel oil	25452	36804	17170	3878	4404	4484
Natural gas	300626	355008	312217	317032	328579	369380
Fuelwood	735	637	644	651	658	665
Total input	481823	613448	592176	589569	656375	776981
Total production	191498	240408	230198	236434	253033	302220
Estimated efficiency (% of production to input)	40	39	39	40	39	39

Statistics on electricity

Wallis and Futuna Is.

Item	2009	2010	2011	2012	2013	2014
Production, trade and consumption	**Million kilowatt-hours**					
Total main activity and autoproducer	20	20	20	19	19	19
Combustible fuels	20	20	20	19	19	19
Hydro	..	..	..	..	..	..
Nuclear	..	..	..	..	..	..
Other	..	..	..	..	..	..
Main activity	20	20	20	19	19	19
Combustible fuels	20	20	20	19	19	19
Hydro	..	..	..	..	..	..
Nuclear	..	..	..	..	..	..
Other	..	..	..	..	..	..
Own use in electricity, CHP and heat plants	3	3	3	2	2	2
Net production	18	17	17	17	17	16
Imports	..	..	..	..	..	..
Exports	..	..	..	..	..	..
Losses	*1	*1	*1	*1	*1	*1
Consumption	*17	*16	*16	*16	*15	*16
Energy industries own use	..	..	..	..	..	..
By industry and construction	..	..	..	..	..	..
By transport	..	..	..	..	..	..
By households and other cons.	*17	*16	*16	*16	*15	*16
Net installed capacity	**Thousand kilowatts**					
Total main activity and autoproducer	9	11	11	11	*11	*11
Combustible fuels	9	11	11	11	*11	*11
Hydro	0	0	0	0	0	0
Nuclear	..	..	..	..	..	..
Other	0	0	0	0	0	0
Main activity	9	11	11	11	*11	*11
Combustible fuels	9	11	11	11	*11	*11
Hydro	0	0	0	0	0	0
Nuclear	..	..	..	..	..	..
Other	0	0	0	0	0	0
Combustible fuel input	**Terajoules**					
Gas-diesel oil	*242	*243	*243	*233	*228	*225
Total input	*242	*243	*243	*233	*228	*225
Total production	72	71	71	68	68	67
Estimated efficiency (% of production to input)	30	29	29	29	30	30

Statistics on electricity

Yemen

Item	2009	2010	2011	2012	2013	2014
Production, trade and consumption	**Million kilowatt-hours**				♦	
Total main activity and autoproducer	6750	7755	6204	7073	8501	7646
Combustible fuels	6750	7755	6204	7073	8501	7646
Hydro	..	..	..	..	..	..
Nuclear	..	..	..	..	..	..
Other	..	..	..	..	..	..
Main activity	6141	7305	5935	6727	8256	7406
Combustible fuels	6141	7305	5935	6727	8256	7406
Hydro	..	..	..	..	..	..
Nuclear	..	..	..	..	..	..
Other	..	..	..	..	..	..
Own use in electricity, CHP and heat plants	466	865	539	580	1341	1206
Net production	6284	6890	5665	6493	7160	6440
Imports	..	..	..	..	..	..
Exports	..	..	..	..	..	..
Losses	1639	1854	1590	2330	2190	1970
Consumption	4645	5036	4075	4163	4970	4470
Energy industries own use	..	..	..	..	..	..
By industry and construction	49	54	148	148	203	161
By transport	..	..	..	..	..	..
By households and other cons.	4596	4982	3927	4015	4767	4309
Net installed capacity	**Thousand kilowatts**					
Total main activity and autoproducer	1334	1334	1535	1535	1535	1519
Combustible fuels	1334	1334	1535	1535	1535	1519
Hydro	..	..	..	..	..	..
Nuclear	..	..	..	..	..	..
Other	..	..	..	..	..	..
Main activity	1216	1216	1417	1417	1417	1402
Combustible fuels	1216	1216	1417	1417	1417	1402
Hydro	..	..	..	..	..	..
Nuclear	..	..	..	..	..	..
Other	..	..	..	..	..	..
Combustible fuel input	**Terajoules**					
Gas-diesel oil	32422	20382	20898	24682	30874	25112
Fuel oil	41248	42501	36804	32603	30017	24280
Natural gas	..	27316	17880	28924	34604	37478
Total input	73670	90199	75582	86209	95495	86870
Total production	24300	27918	22334	25463	30604	27526
Estimated efficiency (% of production to input)	33	31	30	30	32	32

Statistics on electricity

Zambia

Item	2009	2010	2011	2012	2013	2014
Production, trade and consumption	**Million kilowatt-hours**					
Total main activity and autoproducer	9893	10448	11498	12368	13300	14452
Combustible fuels	12	13	15	18	19	410
Hydro	9881	10435	11483	12350	13281	14042
Nuclear	..	..	..	..	..	..
Other	..	..	..	..	..	..
Main activity	9893	10448	11498	12368	13300	14452
Combustible fuels	12	13	15	18	19	410
Hydro	9881	10435	11483	12350	13281	14042
Nuclear	..	..	..	..	..	..
Other	..	..	..	..	..	..
Own use in electricity, CHP and heat plants	232	246	288	287	305	328
Net production	9661	10202	11210	12081	12995	14124
Imports	10	13	120	163	73	13
Exports	589	578	29	980	1083	1256
Losses	1802	1848	2730	945	1139	2162
Consumption	7280	7789	8571	10319	10846	10719
Energy industries own use	..	..	..	..	..	..
By industry and construction	3780	4086	4497	6092	6379	6429
By transport	20	21	23	24	28	31
By households and other cons.	3480	3682	4051	4203	4439	4259
Net installed capacity	**Thousand kilowatts**					
Total main activity and autoproducer	1726	1888	1959	1959	2038	2452
Combustible fuels	80	80	91	91	141	195
Hydro	1646	1808	1868	1868	1897	2257
Nuclear	..	..	..	..	..	..
Other	..	..	..	..	..	..
Main activity	1726	1888	1959	1959	2038	2452
Combustible fuels	80	80	91	91	141	195
Hydro	1646	1808	1868	1868	1897	2257
Nuclear	..	..	..	..	..	..
Other	..	..	..	..	..	..
Combustible fuel input	**Terajoules**					
Gas-diesel oil	301	344	430	430	430	430
Fuel oil	0	0	40	0	0	2990
Total input	301	344	470	430	430	3420
Total production	43	47	54	65	68	1476
Estimated efficiency (% of production to input)	14	14	11	15	16	43

Statistics on electricity

Zimbabwe

Item	2009	2010	2011	2012	2013	2014
Production, trade and consumption	**Million kilowatt-hours**					
Total main activity and autoproducer	7291	8603	9177	9149	9499	10023
Combustible fuels	1833	2840	3975	3761	4517	4592
Hydro	5458	5763	5202	5387	4982	5431
Nuclear	..	..	..	..	..	..
Other	..	..	..	..	..	..
Main activity	7187	8473	9019	8963	9315	9863
Combustible fuels	1729	2710	3817	3575	4333	4432
Hydro	5458	5763	5202	5387	4982	5431
Nuclear	..	..	..	..	..	..
Other	..	..	..	..	..	..
Own use in electricity, CHP and heat plants	282	273	122	99	112	184
Net production	7009	8330	9055	9050	9387	9839
Imports	1882	1682	1579	1076	1722	1127
Exports	1046	694	988	701	1189	1226
Losses	1234	1693	1613	1767	1702	1648
Consumption	7052	7368	8043	7831	8285	8238
Energy industries own use	..	..	..	..	..	..
By industry and construction	2206	3112	3343	3084	3115	3388
By transport	..	..	..	..	..	..
By households and other cons.	4846	4256	4700	4747	5170	4850
Net installed capacity	**Thousand kilowatts**					
Total main activity and autoproducer	2105	2112	*2112	*2112	*2112	1992
Combustible fuels	1355	1355	*1355	*1355	*1355	1235
Hydro	750	757	757	757	757	757
Nuclear	..	..	..	..	..	..
Other	..	..	..	..	..	..
Main activity	2060	2067	*2067	*2067	*2067	1947
Combustible fuels	1310	1310	*1310	*1310	*1310	1190
Hydro	750	757	757	757	757	757
Nuclear	..	..	..	..	..	..
Other	..	..	..	..	..	..
Combustible fuel input	**Terajoules**					
Hard Coal	28211	42816	61848	57474	69596	70325
Gas-diesel oil	301	344	645	731	817	731
Coke-oven gas	0	0	0	..	..	..
Fuelwood	*29	*23	*16	*9	..	..
Bagasse	6184	4924	3665	2405	2390	2074
Total input	34725	48107	66174	60619	72803	73130
Total production	6599	10224	14311	13541	16261	16531
Estimated efficiency (% of production to input)	19	21	22	22	22	23

229

www.ingramcontent.com/pod-product-compliance
Lightning Source LLC
Chambersburg PA
CBHW081433270326
41932CB00019B/3185